# Lecture Notes in Biomathematics

Managing Editor: S. Levin

## 77

A. Hastings (Ed.)

# Community Ecology

A Workshop held at Davis, CA, April 1986

Springer-Verlag

Berlin Heidelberg GmbH

**Editor**

Alan Hastings
Department of Mathematics and Division of Environmental Studies
University of California
Davis, CA 95616, USA

Mathematics Subject Classification (1980): 92A17, 92-06, 92A15

ISBN 978-3-540-50398-9      ISBN 978-3-642-85936-6 (eBook)
DOI 10.1007/978-3-642-85936-6

© Springer-Verlag Berlin Heidelberg 1988
Originally published by Springer-Verlag Berlin Heidelberg in 1988
2146/3140-543210

# INTRODUCTION

This book presents the proceedings of a workshop on community ecology organized at Davis, in April, 1986, sponsored by the Sloan Foundation. There have been several recent symposia on community ecology (Strong et. al., 1984, Diamond and Case, 1987) which have covered a wide range of topics. The goal of the workshop at Davis was more narrow: *to explore the role of scale in developing a theoretical approach to understanding communities.*

There are a number of aspects of scale that enter into attempts to understand ecological communities. One of the most basic is *organizational scale.* Should community ecology proceed by building up from population biology? This question and its ramifications are stressed throughout the book and explored in the first chapter by Simon Levin.

Notions of scale have long been important in understanding physical systems. Thus, in understanding the interactions of organisms with their physical environment, questions of scale become paramount. These more physical questions illustrate the role scale plays in understanding ecology, and are discussed in chapter two by Akira Okubo.

Other questions of organizational scale include how many details, such as genetics or age structure, need to be included in the models of population biology? A related question is: what is the appropriate spatial scale that should be used to understand ecological communities? These two questions of organizational scale are treated in the next two chapters, by Hastings; and Kareiva and Anderson. The role of temporal scale was discussed explicitly in a presentation at the conference which is not included in the present volume.

These first four chapters set the stage for the chapters in the book that deal explicitly with the basic questions of community ecology—why communities have the numbers, densities and kinds of species that we observe. Four different aspects of this question are covered by Chesson, Cohen, Pimm and Yodzis. All of these authors explore the role that scale plays in a particular organizational scheme. An important theme that emerges is that the degree to which the details of organization play a role in understanding the questions of community ecology strongly depends on the detail of the answers sought. Thus, the chapter by Yodzis emphasizes the difficulties in understanding dynamics of even model communities, while the chapter by Cohen emphasizes the role that simple models can play in understanding the numbers of different kinds of species, without considering dynamics. Pimm also discusses the role of models that do not stress dynamics in understanding the numbers and kinds of species. General statements about the role of stochasticity can be made by Chesson within the context of simple models. Chesson explores fully a theme introduced in the first chapter concerning the role played by variability in the choice of an appropriate spatial and temporal scale.

A general conclusion that emerged from the conference was that scale was a valuable concept for organizing theoretical approaches for understanding basic questions in ecology. Determination of the right scale, especially organizational scale, to use when answering a particular question is not easy. Furthermore, there may not be general agreement on what scale is appropriate. However, a choice of scale is almost always a first step in the understanding of ecological processes in nature. This choice of scale is necessary for the experimentalist, the field biologist, and the theoretician.

The conference, and this volume, will have succeeded if some aspects and consequences of the determination of an appropriate scale for ecological investigations have been demonstrated. This volume is certainly not a final answer to questions about scale in ecology. Much more work is needed in ecology by both theoreticians and experimentalists to determine scales of investigation that will prove fruitful in increasing our understanding of nature.

I thank the Sloan Foundation for providing the support for the workshop, the participants who helped make the workshop a success, and Betty McEuen, Patricia Mielke, and Gail Evans for doing such a fine job in producing the final typed version.

## LITERATURE CITED

Diamond, J. and T.J. Case (eds.) 1987. Community Ecology. Harper & Row, New York.
Strong, D., D. Simberloff, L.G. Abele, and A.B. Thistle (eds) 1984. Ecological Communities: Conceptual Issues and the Evidence. Princeton University Press, Princeton, N.J.

# List of Authors

Peter Chesson
> Department of Zoology
> Ohio State University
> 1735 Neil Avenue
> Columbus, Ohio 43210

Joel E. Cohen
> Rockefeller University
> 1230 York Avenue
> New York, New York 10021–6399

Alan Hastings
> Department of Mathematics and Division of Environmental Studies
> University of California
> Davis, CA 95616

Peter Kareiva and M. Andersen
> Department of Zoology
> University of Washington
> Seattle, Washington 98195

Simon Levin
> Ecosystem Research Center and Ecology and Systematics
> Cornell University
> Ithaca, New York 14853

Akira Okubo
> Marine Sciences Research Center
> State University of New York
> Stony Brook, New York 11794–5000
> and
> Ecosystems Research Center
> Cornell University
> Ithaca, New York 14853

Stuart Pimm
> Department of Zoology and Graduate Program in Ecology
> The University of Tennessee
> Knoxville, Tennessee 37996

Peter Yodzis
> Department of Zoology
> University of Guelph
> Guelph, Ontario
> N1G 2W1 Canada

# Table of Contents

CHAPTER 1

Pattern, Scale, and Variability:
An Ecological Perspective

Simon A. Levin
Ecosystems Research Center, and
Section of Ecology and Systematics
Cornell University
Ithaca, NY

I.    INTRODUCTION

One of the fundamental challenges of ecological science is to blend population and community theory, to examine the relationships among phenomena occurring on different scales and the dynamic processes underlying the emergence of pattern. It is a challenge incompletely met; yet community ecology, in its search for integration, is leagues ahead of ecosystems ecology. There, the need and desire for synthesis are at least as great, but the gap separating the subject from population biology remains virgin territory. In each of these quests, reductionistic and holistic approaches must be wedded; in each, the goals are to understand system structure and function in relation to the dynamics at lower levels of organization, and to understand how changes at higher levels may filter down to influence lower levels.

II.   PATTERN

The search to understand any complex system is a search for pattern, for the reduction of complexity to a few simple rules, principles to abstract the signal from the noise. As oceanographers long have recognized, pattern can be found at any level of investigation; and like the sound of the tree falling in the forest, community pattern makes little sense without consideration of the observer. Nietsche (1901) said "There are no facts, only interpretations." Much of the literature on ecological pattern emphasizes equilibrium and homogeneity, reflecting a perspective shaped by historical tradition. When we examine the system in other ways, we find new patterns whose importance is obscured by the classical approach.

In the early twentieth century, as the attention of ecologists turned to community organization, Gleason's emphasis on individualistic and stochastic considerations lost out to Clements' more holistic notion of the climax stable state, and his perception of the community as a superorganism whose characteristics were determined by the local properties of the physical environment (McIntosh 1985). The mathematical theory that emerged from this approach emphasized equilibrium, constancy, homogeneity, stability, and predictability.

A broader perspective, however, makes clear that these attributes are not absolutes, but vary in degree depending on the scale of observation. Systems develop simultaneously on many different scales. On any one scale, one may regard some variables as changing so slowly that in effect they are constant, or so rapidly that only their statistical properties are relevant. But the situation is much more complicated than that, and recognition of the interrelationships among scales is one of the fundamental steps in understanding the development of structure and pattern.

III.   THOUGHTS ON THE DEVELOPMENT OF PATTERN

How does pattern form in the absence of a detailed blueprint? Can simple, localized, contextual rules account for the emergence of pattern at more global scales? This is a pervasive problem in biology, in cosmology, in chemistry, in geology, and indeed in almost any branch of knowledge. In developmental biology, in linguistics, and elsewhere, a central question has been how a few basic rules, largely local in nature, reliably can give rise to recognizable entities at higher levels of organization. Turing (1952) showed how symmetry could be broken through local autocatalysis, reinforcing random or otherwise insignificant inhomogeneities. But the breaking of symmetry is just the first step in the development of pattern; without some mechanism to retard its spread on nonlocal scales, that initial inhomogeneity will give rise to a new homogeneous pattern, simply displacing its predecessor.

What is implicit in Turing's original model (see Levin and Segel 1976, 1984) and in alternative models of pattern generation (see Gierer and Meinhardt 1972; Murray and Oster 1984) is that local activation, as expressed in the enhancement of differences, is in opposition to longer–range inhibition that eventually stabilizes pattern and retards the spread of disturbance. The various models proposed for development differ drastically in their underlying mechanisms, but all successful ones have these two basic features: short–range activation and long–range inhibition (Meinhardt 1982). Indeed, the fact that these two characteristics are all that are needed to produce a very wide range of patterns makes clear the impossibility of discovering process from pattern: quite distinct underlying processes can give rise to identical sets of patterns.

Pattern involves the coexistence of different elements or states, and some regularity in their arrangement. In the theory of population genetics, the first ingredient of pattern is expressed as polymorphism: the coexistence of alleles and of distinct genotypes. The simplest case of allelic coexistence, that of balanced polymorphism, arises because of the superiority of the heterozygote; this may be thought of as gene–level frequency dependence favoring the rare allele, since the rare allele (in contrast to the common one) occurs primarily in the heterozygous form. More generally, whether at the genetical or at the ecological level, frequency dependence favoring rare types, whatever its underlying basis, can play the dual role of catalyzing the spread of local inhomogeneities (short–range activation) and retarding its growth when the inhomogeneities are no longer localized (long–range inhibition). This by itself may not be sufficient to constitute pattern, since no obvious regularities in distribution are expressed; but the essential ingredients are present. This frequency dependence, when coupled with a delay in its operation, can lead to periodic dynamics, clear manifestations of temporal pattern, or to more complicated temporal patterns that at least exhibit statistical regularity. The delay can be explicit, in which case temporal pattern can arise even in a single–variable system, or implicit, operating through the interaction of two or more factors (e.g., predator and prey, or different age classes). In the single–variable case, the concept of selection for rare types is replaced by that of a compensatory mechanism that results in a decreasing per capita population growth rate as population density increases.

The role of the delay in the above example is to assure that activation and inhibition are expressed on different time scales, a central feature of temporal pattern. Similar considerations and mechanisms underlie the generation of patterns in space, a problem that has been studied widely in diverse fields. Because geometrically similar patterns are observed whether one is interested in landscape patterns, animal coat markings, chemical mixtures, thin films of fluids heated from below, or a variety of other situations, it is natural to try to abstract those features that are common to those situations and to develop models that ignore inessential detail. The central aspects of the mechanisms underlying spatial pattern development are some set of rules for local growth or kinetics, and some scheme for redistribution of materials or communication among local environments. The most familiar model systems incorporating these features are those for the diffusion and reaction of chemicals, although the standard models extend easily to more general and nonlocal redistribution regimes. Whatever the context, these models lead to similar consequences.

In the discrete (island) version of this model, pattern can arise as a result of the existence of multiple stable states in the underlying dynamics. The presence of multiple stable states means that the local asymptotic dynamics are influenced by small changes in initial conditions, and hence small differences among local environments become exaggerated due to positive feedback. Thus, we have the first ingredient necessary for pattern to arise: a mechanism for breaking symmetry through short–range activation. In the case of population biology, the initial differences that become enhanced may arise from nothing more than the vagaries of colonization episodes, and the phenomenon usually is described as the "founder effect" (Mayr 1942). Longer–range inhibition is provided by the discrete geometry, which places information exchange among patches on a longer time scale than the instantaneous mixing that is assumed to hold within them, introducing a dichotomy of scales both in space and in time.

In spatially continuous environments, the stabilization mechanism associated with long–range inhibition is lacking, and for the simplest environments no stable non–uniform patterns can result. However, as Matano (1979) has shown, this result, which holds true in convex environments, breaks down in complex geometries that, by forcing materials to flow through bottlenecks, create environments that are quasi–discrete (Fig. 1). Similarly, it need not hold even in convex regions if the diffusion coefficients are spatially non–uniform; such nonuniformity may be determined extrinsically, or may arise through dependence on the local state. In the latter case, an initially homogeneous diffusion regime can become heterogeneous as a result of symmetry–breaking through local activation, and this then may give rise to the quasi–discrete environment necessary to provide longer–range activation (Levin 1979; Fife and Peletier 1980).

Fig. 1:      A geometry that can support non–uniform spatial patterns through the existence of multiple stable states. The key is the existence of bottlenecks.

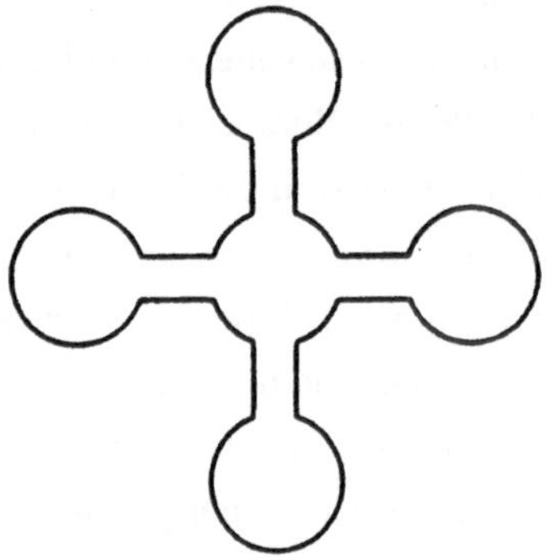

A more explicit way to get short–range activation and long–range inhibition is to assume that there are two separate agents, e.g., chemical morphogens, that specifically fill the roles of activators and inhibitors. In the model of Turing, and the related work of Gierer and Meinhardt (1972), Murray (1981), and others, one assumes that the system has two components: an activator species, whose diffusion is spatially limited, and an inhibitor that diffuses over broader scales. Because symmetry–breaking depends in this case upon the differences in diffusion rates, the phenomenon has been called diffusive instability; the resultant nonuniform pattern is sometimes called a dissipative structure (Glansdorff and Prigogine 1971; see also Levin and Segel 1984). Applications to ecological situations, in which a prey species serves as activator and a predator as inhibitor, are discussed in Segel and Jackson (1972), Levin (1974), Levin and Segel (1976, 1984), and Segel and Levin (1976).

In two–dimensional systems, the mechanisms of activation and inhibition need not be so clearly separable that each resides in a particular species. Levin and Segel (1984), in considering the role of apostatic selection (the tendency of predators to concentrate on common prey types) in fostering diversity, show that nonuniform distributions of character types may arise and be maintained. In this system, symmetry is broken in two ways: initial monomorphic assemblages cannot be maintained, as apostasis provides a mechanism favoring rare types. At the other extreme, completely equitable distributions of competing types may become unstable due to the focusing effects of prey (assortative) mating and reproduction. More generally, activation and inhibition can arise in higher dimensional systems through feedback loops involving many species, or can arise even in one dimension when different phenomena are manifest at different spatial or other scales (Levin and Segel 1984).

## IV.  ASYNCHRONIZED LOCAL DISTURBANCE

The above discussion relates to the development of stable patterns, but such considerations leave out an important class of patterns, those that are transient or are dynamic with some underlying regularity, including chaotic and spatio–temporal patterns.

Following A.S. Watt's prescient presidential address (Watt 1947) to the British Ecological Society in 1947, appreciation grew for the importance of variability in space and time as a factor structuring communities, and as a key to coexistence and coevolution.  As Watt's work and a growing body of later work (e.g., Levin and Paine 1974, 1975; Paine and Levin 1981; Pickett and White 1986) have shown, natural biotic and abiotic disturbance recycles limiting resources, developing mosaics of successional change that allow species to subdivide resources temporally.  The explicit incorporation of disturbance, variability, and stochasticity as part of the description of the normative community is thus an imperative if one is to capture the essential nature of such systems.  For many and perhaps most species, local unpredictability globally is the most predictable aspect of these systems (Levin and Paine 1974).

Work examining the importance of gaps and mosaic phenomena has demonstrated the inseparability of the concepts of equilibrium and scale.  As one moves to finer and finer scales of observation, systems become more and more variable over time and space, and the degree of variability changes as a function of the spatial and temporal scales of observation.  Such a realization long has been part of the thinking of oceanographers, who observe patchiness and variability on virtually every scale of investigation.  A major conclusion is that there is no single correct scale of observation, and that the insights one achieves from any investigation are contingent on the choice of scales.  Pattern is neither a property of the system alone nor of the observer, but of an interaction between them.

The importance of scales also becomes apparent from an examination of population models, both in terms of their general dynamic properties and in terms of their applicability to real populations.  Much recent mathematical work has demonstrated that even the simplest models of populations can exhibit oscillatory and even chaotic behavior; and that, as a consequence, it is impossible to predict accurately the precise dynamics of populations governed by such equations (e.g., May 1974).

To some extent, such investigations render moot the classical debate over whether populations are controlled by density–dependent or density–independent factors. Close to the theoretical equilibrium, the dynamics of such populations may be indistinguishable from those of appropriately chosen stochastic density–independent models; near the equilibrium, density dependence is very weak, and will be obscured by any overriding density–independent variation. On the other hand, far from equilibrium, density–dependent factors assume more importance because the nonlinearities are stronger. Thus, density dependence is the primary mechanism constraining major excursions in population density and keeping populations within bounds; but within those bounds, density–independent phenomena predominate. Concepts of stability that rely on asymptotic return to an equilibrium state are seen to be irrelevant on many scales of interest, and more general concepts such as boundedness and resiliency replace them (Levin 1987).

The increasing recognition that ecological systems are dynamically changing spatiotemporal mosaics has spurred interest in the development of measures that allow comparisons of the importance of disturbance and patchiness across systems and across scales. Hastings et al. (1982) suggest that one approach is to examine the cumulative frequency distribution of patches of various sizes. Their investigation of patch distributions for various successional classes, based both on field data and on the output of simulation models, produces a hyperbolic form (above some threshold patch size) for the cumulative distribution of patch area greater than a given amount. Thus, one has a distribution of the form

$$\text{prob } (A > a) = (\text{const}) \times a^{-B}, \tag{1}$$

where A is patch area for a given successional class. Hastings et al. (1982) transform B by the relation H=2–2B (following Mandelbrot 1977; see also Mandelbrot 1983) to produce a measure that they term "the fractal exponent ... of successional stage." B typically is larger early in succession.

The measure described above is a static one, a snapshot of the system at a particular point in time. As such, it joins a distinguished set of measures of patchiness that community ecologists have used for a long time (see, for example, Greig–Smith 1964, Southwood 1978). But the importance of system dynamics is lost in such measures, and thus there is a need for approaches that look across time as well as across space. In oceanography, the Stommel diagram (Stommel 1963, 1965; Haury et al. 1977) is one means for representing the variability of a system both in space and in time; in geostatistics, various schemes for stratified random sampling achieve the same objective (Bras and Rodriguez–Iturbé 1985). The application of such approaches to ecological systems holds the potential for producing fundamentally new perspectives on these systems, ones that emphasize the changes in the perception of processes across different spatial and temporal scales. Ultimately, these methods can be extended to the consideration of phenomena across organizational scales, and give us powerful new tools for understanding systems.

With my colleague Linda Buttel, I have begun the analysis of successional systems by using this methodology, building on a general successional model that can be tailored to forests, to grasslands, or even to intertidal communities (Levin and Buttel 1986). We have developed a model incorporating disturbance, colonization, and local competition, and investigated its dynamics on the Cornell PSF Supercomputer. In this approach, disturbances of various sizes are superimposed on a grid composed of 10,000 cells, according to a set of stochastic rules that depend on the local states of the system. Disturbances are centered in particular cells, and their size and frequency distribution is conditional upon the current status of the cell (for example, late successional cells are more likely to give rise to larger disturbances through their effects upon neighboring cells); the disturbance then is allowed to radiate outward to adjacent cells. In one version of the model, edge–related disturbances are incorporated; that is, in analogy with systems such as the balsam fir forests studied by Sprugel (1976), trees on exposed edges of disturbances are more susceptible to damage than are more protected trees.

Once a gap is formed, that space is available for recolonization. We assume that colonization comes from a pool, and that different species have different probabilities per unit time of arriving at a site. It is straightforward, although computationally more complicated, to extend the colonization model to include nearest neighbor effects. Competitive ability is assumed to be inversely related to probability of arriving at a site. In the simplest version of the model, a site is occupied by a single individual, selected randomly from among those in the highest competitive class arriving at the site. In more complicated versions, a local growth simulator apportions the local resource (space within a cell) according to a set of rules that allows local coexistence and that implicitly incorporates the time delays that are associated with local competitive displacement.

It is clear that, in this model, the observed temporal variability of any state variable will be a function of the scale of observation. In particular, if one averages a particular measure, such as the percent occupancy of space by a particular species, over a square block of n cells, then the expected temporal variance of that average $(y_n = (x_1 + x_2 + \cdots + x_n)/n)$ is given by

$$s_n^2 = E((y_n - E(y))^2 = E((y_n - E(x))^2) = \sigma^2/n + ((n-1)/n)\text{cov}_n, \qquad (2)$$

where $\sigma^2$ is the variance of x and $\text{cov}_n$ is the covariance of the values of x for two points drawn at random from the square block of n cells. The difficulty is that $\text{cov}_n$ depends on n in a complicated way, and thus it is difficult to derive analytically the relationship between $s^2$ and n.

Clearly, were there no spatial correlation, the variance would fall off inversely with n. More generally, the relationship between $S_n^2$ and n depends on the relationship between $\text{cov}_n$ and n, a relationship that may be very complicated. The empirically derived relationship is somewhat surprising: for every measure that we examined, the relationship between the variance and n was remarkably well approximated on a log–log plot by a linear model (see, for example, Fig. 2). That is, our examination of the temporal variability of nested spatial averages led to the discovery of hyperbolic relationships between variance and scale. For example, for each species examined, the temporal variance of the n–point spatial average (the spatial average for n equally spaced and symmetrically arrayed points) approximately satisfied the relationship

$$S_n^2 = (\text{const})x\, n^{-z},$$

where the exponent z is a measure that reflects the degree of spatial correlation. In general, z varies with successional stage, and is closest to unity for those stages where small–scale patchiness is most important. The deviation from unity is a measure of the spatial scale of disturbance and recovery.

The significance of the above relationship is that, over a broad range of spatial scales, log of the variance of a spatial average is related approximately linearly to the logarithm of the area sampled. Because attention is focused on the slope of this line, the expected value should be independent of the mesh size used; that is, it should be independent of the scale of investigation, facilitating comparisons across systems and across scales. The model on which this relationship is based is a fairly simple one, but the existence of the relationship is very suggestive: if the system is structured according to a single underlying dynamic, then although the observed variance in general will be a function of scale, the slope of the log of variance versus the log of area may represent an invariant, one that is relatively independent of the scale of investigation. If significantly different slopes are found in the investigation of different systems, or of the same system studied on different scales (as we have found for more complicated models), then this suggests that different mechanisms or rates apply in determining those different system structures. More generally, even if a simple relationship hadn't emerged from our investigations, the results would emphasize the importance and value of quantifying the relationship between variability and scale as opposed simply to measuring variance at a single scale.

## V.    SUMMARY

The recognition of what organizes and characterizes a system is a recognition of the manifold patterns the system exhibits. But pattern manifests itself differently on different scales, and the description of system pattern is interwoven with the observer's choice of scales of interest. On any scale, pattern arises from the interplay between order and disorder, between mechanisms upsetting the monotony of homogeneity and those maintaining heterogeneity against the forces of conformity.

Fig. 2: The relationship between variability and scale. See text for discussion.

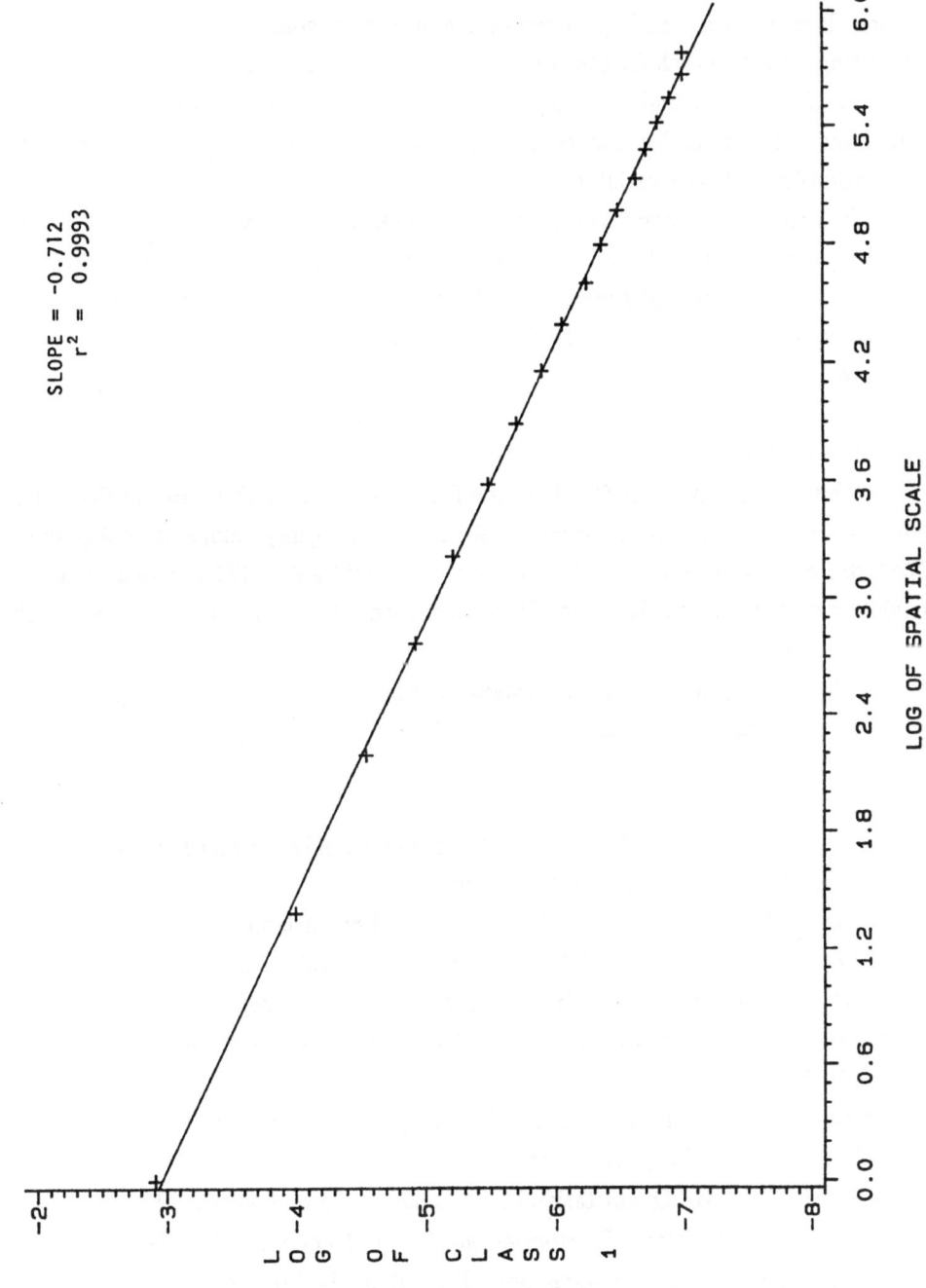

That pattern forms on multiple scales makes evident that focusing on only one scale will give a distorted picture of a system, a single frame in a multi–dimensional motion picture. It emphasizes the importance of examining phenomena across scales rather than conducting a futile search for the true scale of the system. Our study of a model of successional dynamics, and our analysis of the dynamics of that system over a variety of spatial and temporal scales, has led us to discover a number of surprising insights concerning the relationship between variability and scale.

It is clear that these investigations just scratch the surface of what can be learned from examining systems on multiple scales. The escape from single scale studies will provide us with fundamentally new perspectives on the hierarchical dynamics of ecosystems, and may bring some basis for order to the cross–system examination of community and ecosystem structure.

## ACKNOWLEDGMENTS

This is publication ERC–136 of the Ecosystems Research Center of Cornell University, and was supported by U.S. Environmental Protection Agency Cooperative Agreement CR812685 and National Science Foundation Grant DMS–8406472 to Simon A. Levin. The opinions expressed are solely those of the author, and do not necessarily represent the views of the sponsoring agencies.

The author gratefully acknowledges the contributions of Lee A. Segel and Linda Buttel to the development of the ideas expressed.

## REFERENCES

Bras, R.L. and I. Rodriguez–Iturbé. 1985. Random Functions and Hydrology. 559 + xv pp. Addison–Wesley, Reading, Massachusetts.

Fife, P. and L.A. Peletier. 1980. Clines introduced by variable migration. pp. 276–278. In W. Jäger, H. Rost, and P. Tautu (eds.). Biological Growth and Spread. Lecture Notes in Biomathematics 38, Springer–Verlag, Heidelberg.

Gierer, A. and H. Meinhardt. 1972. A theory of biological pattern formation. Kybernetik 12:30–9.

Glansdorff, P. and I. Prigogine. 1971. Thermodynamic Theory of Structure, Stability, and Fluctuations. Wiley, New York.

Greig–Smith, P. 1964. Quantitative Plant Ecology. Second edition. Butterworths, London.

Hastings, H.M., R. Pekelney, R. Monticciolo, D. vun Kannon, and D. DelMonte. 1982. Time scales, persistence, and patchiness. Biosystems 15:281–289.

Haury, L.R., J.A. McGowan, and P.H. Wiebe. 1977. Patterns and processes in the time–space scales of plankton distributions. pp. 277–328. In J.H. Steele (ed.). Spatial Pattern in Plankton Communities. Plenum Press, New York.

Levin, S.A. 1974. Dispersion and population interactions. Amer. Natur. 108:207–228.

Levin, S.A. 1979. Non–uniform stable solutions to reaction–diffusion equations: Applications to ecological pattern formation. pp. 210–222. In H. Haken (ed.). Pattern Formation by Dynamic Systems and Pattern Recognition. Springer–Verlag, Berlin.

Levin, S.A. 1987. Scale and predictability in ecological modeling. In Proc., Workshop on Applied Control Theory to Renewable Resource Management, Honolulu (in press).

Levin, S.A. and L. Buttel. 1986. Measures of patchiness in ecological systems. Ecosystems Research Center Report No. ERC–130, Cornell University, Ithaca, New York.

Levin, S.A. and R.T. Paine. 1974. Disturbance, patch formation, and community structure. In Proc. Nat. Acad. Sci. USA 71:2744–47.

Levin, S.A. and R.T. Paine. 1975. The role of disturbance in models of community structure. In Ecosystem Analysis and Prediction, S.A. Levin, (ed.), pp. 56–67. Proceedings of a Conference on Ecosystems, Alta, Utah. SIAM–SIMS, Philadelphia, Pennsylvania.

Levin, S.A. and L.A. Segel. 1976. Hypothesis for origin of planktonic patchiness. Nature 256:659.

Levin, S.A. and L.A. Segel. 1984. Pattern generation in space and aspect. SIAM Review 27:45–67.

Mandelbrot, B.B. 1977. Fractals: Form, Chance, and Dimension. W.H. Freeman & Co., San Francisco.

Mandelbrot, B.B. 1983. The Fractal Geometry of Nature. W.H. Freeman & Co., 468 pp.

Matano, H. 1979. Asymptotic behavior and stability of semi–linear diffusion equations. Publ. Res. Inst. Math. Sci., Kyoto. 15:401–51.

May, R.M. 1974. Biological populations with non–overlapping generations: stable points, stable cycles, and chaos. J. Theor. Biol. 49:511–524.

Mayr, E. 1942. Systematics and the Origin of Species. Columbia University Press, New York.

McIntosh, R.P. 1985. The Background of Ecology: Concept and Theory. Cambridge University Press, Cambridge, England.

Meinhardt, H. 1982. Models of Biological Pattern Formation. Academic Press, New York.

Murray, J.D. 1981. A prepattern formation mechanism for animal coat markings. J. Theor. Biol. 88:161–199.

Murray, J.D. and Oster, G.F. 1984. Cell traction models for generating pattern and form in morphogenesis. J. Math. Biology 19:265–279.

Nietzsche, F. 1901. Wille Zur Macht. Edited by E. Förster–Nietzsche. Kröner, Leipzig.

Paine, R.T. and S.A. Levin. 1981. Intertidal landscapes: Disturbance and the dynamics of pattern. Ecol. Monogr. 51:145–178.

Pickett, S.T.A. and P.S. White. 1986. Natural Disturbance and Patch Dynamics. Academic Press, Orlando, Florida.

Segel, L.A. and J.L. Jackson. 1972. Dissipative structure: An explanation and an ecological example. J. Theor. Biol. 37:545–559.

Segel, L.A. and S.A. Levin. 1976. Applications of nonlinear stability theory to the study of the effects of dispersion on predator–prey interactions. pp. 123–52. In R. Piccirelli (ed.). Selected Topics in Statistical Mechanics and Biophysics. Conference Proceedings Number 27, American Institute of Physics, New York.

Southwood, T.R.E. 1978. Ecological Methods with Particular Reference to the Study of Insect Populations. Second Edition, Halsted.

Sprugel, D.G. 1976. Dynamic structure of wave–regenerated Abies balsamea forests in the northeastern United States. J. Ecol. 64:889–911.

Stommel, H. 1963. Varieties of oceanographic experience. Science 139:572–576.

Stommel, H. 1965. Some thoughts about planning the Kuroshio Survey. In Proc. Symp. on the Kuroshio, Tokyo, Oct. 29, 1963. Oceanogr. Soc. Japan and UNESCO.

Turing, A.M. 1952. The chemical basis of morphogenesis. Phil. Trans. Roy. Soc. B. 237:37–72.

Watt, A.S. 1947. Pattern and process in the plant community. J. Ecol. 35:1–22. Proceedings of a Conference at Davis in April, 1986

CHAPTER 2

Planktonic Micro—Communities in the Sea:
biofluid mechanical view

Akira Okubo

Marine Sciences Research Center
State University of New York
Stony Brook, NY 11794—5000
and
Ecosystems Research Center
Cornell University
Ithaca, NY 14853—2701

I.    INTRODUCTION

It has long been known that marine planktonic organisms are patchily distributed at a variety of spatial scales (Steele, 1976, 1978; Haury et al, 1978). Until recently, however, most studies of plankton patchiness have focused on spatial scales larger than the order of 10 cm (vertical) and of 10 m (horizontal). Recently advances in instrumentation have now made it possible to examine patches smaller than 1 m in horizontal direction (Denman and Mackas, 1978; Platt, 1978). Evidence indicates that microscale patches as small as or smaller than 1 cm in diameter are common and persistent in the turbulent sea (Mitchell, 1988). These patches may have a significant role in ecosystem function. Most importantly, although organisms and nutrients are on average sparsely distributed in oceanic waters, their coassociation in micropatches could be key to energy and nutrient flows (McCarthy and Goldman, 1979).

To understand these microscale associations, several investigators have suggested that we need to look in detail at the behavioral responses of organisms in their potentially patchy fluid media and at the transport process which may either maintain or dissipate heterogeneities in nutrient distribution. This analysis will require careful attention to the fluid flows in the vicinity of planktonic organisms. All questions involve situations in which microscale patchiness and the response of organisms to this patchiness might be key to species interactions.

One problem of interest is how can oligotrophic oceans, traditionally thought of as biological deserts, be biologically productive as has recently be learned. In these cases nitrogenous nutrients are below detection limits even though considerable photosynthetic activity is noted. To resolve this dilemma of "missing matter" in the ocean, McCarthy and Goldman (1979) hypothesized that extremely small—scale nutrient patches either excreted locally by zooplankton or remineralized by bacteria should be continuously utilized by adjacent phytoplankton cells at rates that could maintain the bulk concentration averaged over large water volumes at the level below detection limits.

Another problem of the missing matter dilemma has been proposed by Azam and Ammerman (1984). They noted that bacteria live in the ocean water of low dissolved organic matter concentration, but appear to grow rapidly. This leads to the hypothesis that bacteria cluster around phytoplankton cells or particulate nutrient sources to utilize released organic carbon orders of magnitude above background concentrations.

To address questions or hypotheses such as those above we need to pursue theoretical calculations as well as experiments. This is especially true because experiments are so difficult to perform at these microscales. The needed theoretical analyses must include physical, chemical, and biological processes. Especially important will be a fluid dynamical understanding of how the flow pattern around marine organisms affects their microscale interactions (Jackson, 1986; Okubo, 1987).

II.     FLUID MECHANICS RELEVANT TO MARINE ORGANISMS

For almost all motions in marine ecology we may assume that the environmental flow is incompressible and Newtonian, and that variations in the fluid density can be neglected in so far as they influence inertia. For the movement of marine organisms the effect of the Coriolis force can also be ignored in comparison with either inertial or viscous forces (Okubo, 1987). Thus the equations of motion for the ambient water are:

$$\partial \underset{\sim}{u}/\partial t + (\underset{\sim}{u}\cdot\nabla)\underset{\sim}{u} = \rho_0^{-1}\nabla p + \nu\nabla^2\underset{\sim}{u} + (\rho-\rho_0)/\rho_0\,\underset{\sim}{g} \qquad (1)$$

$$\nabla\cdot\underset{\sim}{u} = 0, \qquad (2)$$

where $\underset{\sim}{u}$ is fluid velocity vector, p is pressure, $\rho$ is fluid density, $\rho_0$ is reference density, g is the acceleration of gravity, $\nu$ is the kinematic viscosity of water. The buoyancy term, which is the last term on the right–hand side of (1), plays no role in the analysis that follows because it can be absorbed into the pressure gradient (Batchelor, 1967; Childress, 1981).

A marine organism is usually considered as an entity separated from the environmental water and thus may be treated as an impenetrable region on whose boundary the fluid has no slip, i.e., no relative velocity at the organisms surface. If the number of organisms is very great and organism dimensions are very small compared with the dimensions of the system of concern, then a seawater–organism suspension can be regarded as a continuum.

The relative importance of the inertial and viscous forces is measured by the Reynolds number expressed by $R_e = U\ell/\nu$, where U and $\ell$ are some representative scales of velocity and length. Okubo and Mitchell (Okubo, 1987) calculated Reynolds numbers associated with movement of marine organisms ranging from bacteria to whales. The relationship between the Reynolds number and the body length of organism $\ell$ is given by

$$R_e = 269 \, \ell^{1.86} \quad (\ell \text{ in cm}) \tag{3}$$

Thus the movement of bacteria and phytoplankton is characteristic of the low Reynolds number flow, where viscous forces dominate inertial forces. The movement of zooplankton, on the other hand, ranges from low to intermediate Reynolds numbers. Zooplankton of length 1 mm experience Reynolds numbers of the order of unity in their swimming behavior.

The knowledge of the flow pattern around a phytoplankton cell enables us to calculate the concentration distribution of a chemical released from the cell. The problem is mathematically treated by the advection, diffusion, and reaction equation of the chemical concentration $C$

$$\partial C / \partial t = - \underset{\sim}{u} \, \underset{\sim}{\nabla} \, C + D \, \nabla^2 \, C + F(C)$$

where $\underset{\sim}{u}$ is advection velocity, $D$ is molecular diffusivity and $F$ is reaction rate.

## III. INTERACTION OF ZOOPLANKTON AND PHYTOPLANKTON IN MICROSCALES.

Recent progress in microcinematographic techniques has allowed us to directly observe small zooplankton (copepods) feeding on even smaller algae (Paffenhöfer et al., 1982; Koehl, 1984; Strickler, 1982; Price et al, 1983; Vanderploeg and Paffenhöfer, 1985). These observations reveal that a copepod is capable of perceiving the presence of an algal particle at a distance (0.1 ~ 1 mm) away, reorienting itself, and adjusting the feeding current so as to draw more efficiently the prey particle toward itself. Chemoreception and mechanoreception have been suggested as the sensory mechanisms underlying particle detection and size selection in copepods. Both of these mechanisms involve the low Reynolds number flow.

Phytoplankton cells are surrounded by an "active space" (Andrews, 1983) within which the concentration of simple sugars, amino acids and other excudates exceeds background levels (see also Section 4). These leaked chemicals are the basis for chemoneception in zooplankton. In particular, the active space may be deformed by the presence of a feeding current, so that an initially spherical isoconcentration surface is elongated in the direction of the flow. The feeding zooplankter may then be able to chemically detect the presence of prey from a distance. For more details consult Andrews (1983). One of the weak points in this chemoreception model is that the threshold concentration of chemicals, zooplankton are assumed to be able to detect seem rather small compared with the concentrations of amino acids required for the behavioral responses in zooplankton (Legier–Visser et al., 1986).

Mechanoreception as a means of prey detection for zooplankton has been proposed by Legier–Visser et al. (1986). This model is based on the pressure field induced by a prey particle. For a spherical particle embedded in a low Reynolds number flow the Stokes solution of the pressure field gives:

$$\Delta p = 3 \eta a U \cos\theta / r^2 \qquad (4)$$

where $\Delta p$ is the pressure difference between flow in the absence of a particle and flow that is deformed by a particle of radius a; $\eta$ is the viscosity of the fluid; $\Theta$ is the angle between the particle and the mechanoreceptor of copepod; r is the distance from the center of the particle to the antenna of copepod measured perpendicular to the antenna; U is the flow velocity far away from the particle.

According to this model, for given flow conditions and a minimum pressure sensitivity of the copepod's mechanoreceptor, the distance of prey detection increases with prey size, reaches a maximum value, and decreases toward zero as the prey size increases. Thus we expect that copepods would have the greatest detection ability in a narrow range of particle size. This maximum detection distance and optimal size of prey are determined in part by the intensity of the feeding current which may, in fact, be altered by the animal.

McCarthy and Goldman (1979) suggested that microscale patches of nutrients excreted by zooplankton should be an important nutrient source of phytoplankton in the oligotrophic ocean. Mathematical modelling of the nutrient plumes or patches depends upon the advection–diffusion–reaction equation for the concentration of nutrient excreted from zooplankton

$$\partial C/\partial t + u\, \partial C/\partial x = D\, \nabla^2 C - V_m PC/(C + K) \qquad (5)$$

where C is nutrient (ammonia) concentration, u is advection velocity in x–direction behind the zooplankter, D is molecular diffusivity, $V_m$ is maximum uptake rate, K is the half–saturation constant, P is the concentration of phytoplankton cells. The assumption of molecular diffusion may be appropriate for the low Reynolds number regime.

An approximate solution of (5) is obtained by linearizing the reaction term when C << K. Then, the steady–state solution of (5) under a continuous excretion of constant rate is given by: where $r^2 = x^2 + y^2 = z^2$, $\lambda = v_m P/K$, and Q is rate of nutrient release per animal. The following numerical values are chosen to correspond to the condition of the oligotrophic ocean: $Q = 10^{-12}$ mol/sec (Ikeda, 1977; Verity, 1985) u = 0.2 ~ 1 cm/sec

$$C(x) = \frac{Q}{4\pi Dr} \exp - [\{(u^2 + 4\lambda D)^{1/2}\, r - ux\}/2D] \qquad (6)$$

(Kirk, 1985; Currie, 1984a,b; Andrews, 1983), $V_m = 10^{-14}$ mol/sec (Currie, 1984a,b), P =

$10^3/cm^3$, K = 0.5 $\mu$ M, D = $10^{-5}$ $cm^2$/sec. $C^*$ (threshold ammonia concentration) = 0.5 $\mu$ M (= 10 times the background concentration).

These values give the maximum extent of the nutrient plume behind a zooplankter as about 20 cm, and the volume containing the nutrient concentration above the threshold approximately as ranging from 0.05 to 0.25 $cm^3$ (with the lower and upper limits corresponding to u = 1 and u = 0.2 cm/sec, respectively.) This calculated effective volume of nutrient plume amounts to only $5 \times 10^{-5}$ to $2.5 \times 10^{-4}$ of the total volume of sea water for a density of zooplankton of one animal per liter. Even for 10 animals per liter the volume is only $5 \times 10^{-4}$ to $2.5 \times 10^{-3}$ of the total. Thus, even if we take into account plumes of nutrients surrounding each zooplankton, nutrients are still remarkably scarce in the waters of open oceans.

If, instead of continuous release, we consider a case where the nutrient excretion occurs once a day with the total amount m = Q x 24 hours, then the solution of the linearized version of (5) is given by:

$$C(c,t) = \frac{m}{8\pi^{3/2}D^{3/2}t^{3/2}} \exp - \{ \frac{(x - ut)^2 + y^2 + z^2}{4Dt} \} \tag{7}$$

After one day the volume containing the nutrient concentration above the threshold is approximately 44.5 $cm^3$, which corresponds to 4.5% of the total volume of sea water for one animal per liter.

Real nutrient excretion in zooplankton must be intermittent. Therefore the effective volume of a nutrient patch should be less than 44.5 $cm^3$ and more than that given for the continuous case. According to McCarthy and Goldman (1979) phytoplankton need to be exposed to a high concentration of ammonia during only about 3% of a doubling period, say one day. The above calculations though rather crude do not seem to support McCarthy and Goldman's hypothesis. Jackson (1980) pointed out that such micropatches of nutrient might dissipate by turbulent diffusion too quickly to be used by phytoplankton. On the other hand, Lehman and Scavia (1982) show both theoretically and experimentally that algae can exploit micropatches of dissolved phosphate produced by zooplankton. Currie (1984a,b) argued that a nonhomengeneous nutrient supply regime should actually decrease phytoplankton growth rates, regardless of the physical characteristics of the patches. The controversy is by no means settled.

In addition to the problem of diffusion the patch encounter rate of algal cells must be studied in detail. In this context Strickler (1984) used high speed microcinematography to show that feeding currents of copepods result in the concentration of algal cells near the animal and that most of these cells appear to survive the encounter. This implies that phytoplankton cells are favorably entrained into the excreted micropatches (McCarthy and Altabet, 1984).

## IV. INTERACTION OF PHYTOPLANKTON AND BACTERIA: MICROZONES

Most marine bacteria suspend freely in the environmental water rather than being attached to suspended particles such as plankton cells. Nonetheless, peaks of bacterial cell numbers in the depth profile often coincide with peaks of phytoplankton cell numbers (Bird and Kalff, 1984; Fuhrman et al., 1980, 1985). In addition bacterial productivity is closely correlated with phytoplankton productivity (Albright, 1977).

Azam and Ammerman (1984) noted that marine bacteria appear to grow rapidly in the environmental water although the background concentration of dissolved organic matter is too low to support significant bacteria growth. They hypothesized that there are microzones around phytoplankton cells which contain dissolved organic carbon at concentrations orders of magnitude above background concentrations. By clustering in the microzones, bacteria could make use of the high concentrations of organic matter needed to support significant growth.

Another observation is that little of the dissolved organic carbon released by phytoplankton is present in natural waters. This implies that the bacteria use this material as fast as it is produced (Herbst and Overbeck, 1978). This may be further support for the hypothesis of bacterial clustering, although it is equally possible that free floating bacteria without clustering utilize the dissolved material in waters as fast as produced.

Phytoplankton cells release from 1% to 70% of their photosynthetic products, up to 80% of which may be readily used by bacteria (Chrost and Faust, 1983; Søndergaad et al. 1985). This suggests that there may be an intimate link between algae and bacteria as a result of the dissolved organic material surrounding phytoplankton. One can calculate how closely bacteria would have to be associated with phytoplankton cells in order to take advantage of the algae's released organic matter.

Mitchell et al. (1985) estimated the size of the microzones on the basis of advection and diffusion equation

$$\partial C/\partial t = -\underset{\sim}{u}\,\underset{\sim}{\nabla}\,C + D\,\nabla^2\,C \qquad (8)$$

where $C$ is the concentration of dissolved organic carbon (DOC), $\underset{\sim}{u}$ is the advection velocity and $D$ is molecular diffusivity. For a steady-state distribution of nutrient around a phytoplankton cell at rest, the radius of the microzone $R^*$ is estimated by

$$R* = Q/4\pi DC* \qquad (9)$$

where $Q$ is the total DOC flux per cell and $C^*$ is the threshold concentration (taken as 10% above the background concentration). Given $Q = 10^{-17}$ mol/sec, $D = 10^{-5}$ cm$^2$/sec and $C^* = 10^{-12}$ mol/cm$^3$, Mitchell et al. calculated

$$R^* \sim 1 \text{ mm} \qquad\qquad (10)$$

Note that (10) also gives the maximum extent of the microzone behind a sinking phytoplankton cell. Solving (8) with $\underset{\sim}{u} = w_s \underset{\sim}{k}$ ($w_s$: settling velocity, $\underset{\sim}{k}$: unit vertical vector), we obtain the concentration of DOC around a sinking cell

$$C(\underset{\sim}{x}) = (Q/4\pi DR) \exp\{-w_s(R-z)/2D\} \qquad\qquad (11)$$

where $R^2 = x^2 + y^2 + z^2$ and the cell is sinking in z–direction.

Interestingly, the value of the microzone $R^*$ is nearly the same as the length scale of Kolmogorov micro–eddies, i.e. the smallest turbulence found in the upper mixed layer of the ocean (Mitchell et al., 1985). However, in the deep waters of the oceans and in the thermocline below the upper mixed layer, low turbulence conditions prevail and the characteristic length scale of the Kolmogorov eddies is of the order of 1 cm. Mitchell et al. suggest that the deep water and in particular "sheets" and "layers" composing the thermocline are the most likely location for microzones.

Yet the degree of clustering of bacteria around phytoplankton cells is still mostly unknown. Azam and Ammerman (1984) hypothesize that bacteria respond to the microzone by means of chemotaxis and motility thus forming bacteria clusters in the vicinity of algal cells. Jackson (1986) has recently developed a computer model which simulates the behavior of marine bacteria around an algal cell using the model of Brown and Berg (1974) for bacteria chemotaxis and using the model of DOC diffusion around a cell. Based on these simulations Jackson (1986) concludes that initially uniform bacteria can aggregate around algal cells of certain size range.toward the cell. Thus, for a 10 $\mu$ radius alga, leakage rates of DOC cause bacterial drift toward the alga, but for a 2.5 $\mu$ radius alga, there is no drift toward the cell even for high leakage rates. This suggests that there is a minimum size of alga below which chemotactic aggregation cannot be effectively used by bacteria.

Jim Mitchell (private communication) has hypothesized that motile marine bacteria can orient themselves toward a sinking phytoplankton cell by gyrotaxis and then swim toward the cell. This mechanism is independent of chemotaxis. Let us examine this hypothesis on the basis of fluid mechanics. Gyrotaxis is a directed locomotion resulting from the orientation of cell's axis by compensating gravitational and viscous torques (Kessler, 1984, 1985a,b). Relative to the center of buoyancy or drag, the center of gravity of a bacterium is displaced due to an asymmetric distribution of heavy organelles in the cell. The force of gravity then exerts a torque so as to orient the cell axis vertically, and the bacterium tends to swim vertically upward. However, when the environmental water has spatial variation in its velocity, the bacterium experiences an additional hydrodynamic torque. At the balance of these two torques (Kessler, 1984),

$$4\pi\,\eta\,a^3\underset{\sim}{\omega} + 4/3\,\pi\,a^3\,\Delta\rho\,\underset{\sim}{L}\times\underset{\sim}{g} = 0, \tag{12}$$

where a is the radius of bacterial cell, $\underset{\sim}{\omega}$ is the vorticity of the environmental water, $\Delta\rho$ is the density difference between bacterial cell and water, and $\underset{\sim}{L}$ is the displacement vector from the center of buoyancy to the center of gravity. Equation (12) specifies the orientation of the cell axis and thus the direction of swimming.

Consider a situation where a phytoplankton cell of radius a is sinking vertically downward with a settling velocity of $u_0$ and at a great distance from the cell, bacteria are swimming vertically upward at a speed of v. This system is equivalent to that of the phytoplankton cell being at rest and the environmental water being moved upward with a speed of $u_0 + v$, in which bacteria are embedded. Application of Stokes' solution to the latter system yields the flow components

$$U_r \text{ (radial)} = \{(u_0 + v) - 3/2\,u_0\,a/r + 1/2\,u_0\,a^3/r^3\}\cos\Theta \tag{13}$$

$$U_\Theta \text{(tangential)} = \{-(u_0 + v) + 3/4\,u_0\,a/r + 1/4\,u_0\,a^3/r^3\}\sin\Theta \tag{14}$$

where the coordinate origin is taken at the center of the phytoplakton cell.

The vorticity of the flow is given by

$$\omega = -3/2\,u_0\,a/r^2\sin\Theta \tag{15}$$

Substitution of (15) into (12) yields:

$$\sin\phi = (9u_0 a/2\Delta\rho\,Lg)\sin\Theta/r^2 \tag{16}$$

where $\phi$ is the angle of deviation of the bacterium cell axis from the vertically upward orientation, and the angle is taken in such a manner that the bacterium tends to orient toward the phytoplankton cell. Thus the angle $\phi$ increases counterclockwise when bacteria are approaching the algal cell to the left of it. Since L is proportional to the size of organism, the gyrotactic orientation is more effective in microorganisms than the larger organisms.

This orienting effect can occur in any vorticity fields, provided the balance between gravity and viscous drag torques is established. For instance, self–focused bacterial streamers (Kessler, 1986) may spontaneously generate in intermittent fully–developed turbulence in the sea. Local fluctuations in cell concentration produce locally higher concentrations of cell and the denser part sinks, thereby generating a fluid velocity field somewhat similar to that generated by a sinking cell of phytoplankton. By gyrotaxis more cells are drawn into this part of the fluid to accumulate. This process is characteristic of a positive feedback. Thus initially uniformly distributed cells tend to break up into downwelling columns (streamers) containing higher than average cell concentrations. Since the micro–scale turbulent eddies in the sea are very intermittent, the space between the micro–eddies should be a domain of temporary quiescence, and there the microorganism streamers may survive for a while before breaking up by the sporadic action of the micro–eddies.

## V.    CONCLUSION

Planktonic organisms in the sea are patchy on almost any temporal and spatial scales that one can imagine. At the smallest end of the scale interactions and behavior of individual organisms play an important role in the community structure. Since the movement of these organisms is often mercifully controlled by the motion of the environmental fluid, we need to have a biofluidmechanical understanding of the behavior of organisms.

The interface between the environmental fluid and organisms is extensive. There is a long list of questions in planktonic community dynamics to which only an interdisciplinary blend of biology and fluid mechanics can provide the answers.

Acknowledgements

This work is a result of stimulating discussions with my colleagues: Jed Fuhrman, George Jackson, Marianne Legier, and Jim Mitchell to name only some.

I thank Alan Hastings for inviting me to the conference on community ecology at Davis and Peter Kareiva for helping me to improve the manuscript.

Contribution No. 599 of the Marine Sciences Research Center, State University of New York at Stony Brook and Contribution No. ERC–172 of the Ecosystems Research Center, Cornell University.

## REFERENCES

Albright, L.J. (1977). Hetereotrophic bacterial dynamics in the lower Fraser River, its estuary and Georgia Strait, British Columbia. Mar. Biol. 39:203–211.
Andrews, J.C. (1983). Deformation of the active space in the low Reynolds number feeding current of calanoid copepods. Can. J. Fish. Aquat. Sci. 40:1293–1302.

Azam, F. and J.W. Ammerman (1984). Cycling of organic matter by bacterioplankton in pelagic marine ecosystems: microenvironmental considerations. In: Flows of Energy and Material in Marine Ecosystems, 345–360, M.J.R. Fasham (ed.), Plenum Press, New York.

Batchelor, G.K. (1967). An Introduction to Fluid Dynamics. Cambridge University Press, London and New York, 615 pp.

Bird, D.F. and J. Kalff (1984). Empirical relationships between bacterial abundance and chlorophyll concentration in fresh and marine waters. Can. J. Fish. Aquat. Sci. 41:1015–1023

Brown, D.A. and H.C. Berg (1974). Temporal stimulation of chemotaxis in Escherichia coli. Proc. Nat. Acad. Sci. USA, 71:1388–1392.

Childress, S. (1981). Mechanics of Swimming and Flying. Cambridge University Press, London and New York, 155 pp.

Chrost, R.H. and M.A. Faust (1983). Organic carbon release by phytoplankton: its composition and utilization by bacterioplankton. J. Plankton Res. 5:477–493.

Currie, D.J. (1984a). Phytoplankton growth and the microscale nutrient patch hypothesis. J. Plankton Res. 6:591–599.

Currie, D.J. (1984b). Microscale nutrient patches: do they matter to the phytoplankton? Limnol. Oceanogr. 29:211–214.

Denman, K.L. and D.L. Mackas (1978). Collection and analysis of underway data and related physical measurements. In: Spatial Pattern in Plankton Communities, 85–110, J.H. Steele (ed.), Plenum Press, New York.

Fuhrman, J.A., J.W. Ammerman and F. Azam (1980). Bacterioplankton is the coastal euphoric zone: distribution, activity, and possible relationships with phytoplankton. Mar. Biol. 60:201–207.

Fuhrman, J.A., R.W. Eppley, A. Hagstrom and F. Azam (1985). Diel variations in bacterioplankton, phytoplankton, and related parameters in the Southern California Blight. Mar. Ecol. Prog. Ser. 27:9–20.

Haury, L.R., J.A. McGowan and P.H. Wiebe (1978). Patterns and processes in the time–space scales of plankton distributions. In: Spatial Pattern in Plankton Communities, 277–327, J.H. Steele (ed.), Plenum Press, New York.

Herbst, v. and Y. Overbeck (1978). Metabolic coupling between alga Osscillatoria redekei and accompanying bacteria. Naturwissenschaften 65:598–608.

Ikeda, T. (1977). The effect of laboratory conditions on the extrapolation of experimental measurements to the ecology of marine zooplankton. 4. Changes in respiration and excretion rates of boreal zooplankton species maintained under fed and starved conditions. Mar Biol. 41:241–252.

Jackson, G.A. (1980). Phytoplankton growth and zooplankton grazing in oligotrophic oceans. Nature 284:439–441.

Jackson, G.A. (1986). Physical and chemical properties of aquatic environments, Proc. Symposium on "Ecology of Microbiological Communities," 108th Ordinary Meeting of the Society for General Microbiology. Univ. St. Andrews, Canada.

Kessler, J.O. (1984). Gyrotactic buoyant convection and spontaneous pattern formation in algal cell cultures. In: Nonequilibrium Cooperative Phenomena in Physics and Related Fields, 241–248, M.G. Velarde (ed.), Plenum Press, New York.

Kessler, J.O. (1985a). Co–operative and concentrative phenomena of swimming micro–organisms. Contemporary Physics, 26:147–166.

Kessler, J.O. (1985b). Hydrodynamic focusing of motile algal cells. Nature 313:218–220.

Kessler, J.O. (1986). Individual and collective fluid dynamics of swimming cells. J. Fluid Mech. 173: 191–205.

Kirk, K.L. (1985). Water flows produced by Daphnia and Diaptomus: implications for prey selection by mechanosensory predators. Limnol. Oceanogr. 30:679–686.

Koehl, M.A.R. (1984). Mechanisms of particle capture by copepods at low Reynolds numbers: possible modes of selective feeding. In: Trophic Interactions within Aquatic Ecosystems, 135–164, D.G. Meyers and J.R. Strickler (ed.), AAAS Selected Symposium 85, Westview Press, Boulder, Colorado.

Legier–Visser, M., J.G. Mitchell, A. Okubo and J.A. Fuhrman (1986). Mechanoreception in calanoid copepods: a mechanism for prey detection. Mar. Biol. 90:529–535.

Lehman, J.T. and D. Scaria (1982). Microscale nutrient patches produced by zooplankton. Proc. Nat. Acad. Sci. USA 79:5001–5005.

McCarthy, J.J. and J.C. Goldman (1979). Nitrogeneous nutrition of marine phytoplankton in nutrient–depleted waters. Science 203:670–672.

McCarthy, J.J. and M.A. Altabet (1984). Patchiness in nutrient supply: implications for phytoplankton ecology. In: Trophic Interactions within Aquatic Ecosystems, 29–47, D.G. Meyers and J.R. Strickler (ed.), Westview Press, Boulder, Colorado.

Mitchell, J.G. (1988). A new mechanism for generating plankton heterogeneity on small scales. Ph.D. dissertation, State University of New York at Stony Brook.

Mitchell, J.G., A. Okubo and J.A. Fuhrman (1985). Microzones surrounding phytoplankton form the basis for a stratified marine microbial ecosystem. Nature 316:58–59.

Okubo, A. (1987). Fantastic voyage into the deep: marine biofluid mechanics. In: Mathematical Topics in Population Biology, Morphogenesis and Neurosciences, 32–47, E. Teramoto and M. Yamaguti (ed.), Springer–Verlag.

Paffenhöfer, G.A, J.R. Strickler and M. Alcaraz (1982). Suspension–feeding by herbivorous calanoid copepods: a cinematographic study. Mar. Biol. 67:193–199.

Platt, T. (1978). Spectral analysis of spatial structure in phytoplankton populations. In: Spatial Pattern in Plankton Communities, 73–84, J.H. Steele (ed.), Plenum Press, New York.

Price, H.J., G.A. Paffenhöfer and J.R. Strickler (1983). Modes of cell capture in calanoid copepods. Limnol. Oceanogr. 28:116–123,

Søndergaad, M., B. Riemann and N.O.G. Jorgensen (1985). Extracellular organic carbon (EOC) released by phytoplankton and bacterial production. Oikos 45:323–332.

Steele, J.H. (1976). Patchiness. In: Ecology of the Sea, 95–115, D.H. Cushing and J.J. Walsh (ed.), W.B. Saunders Co., Philadelphia.

Steele, J.H. (1978). Some comments on plankton patches. In: Spatial Pattern in Plankton Communities, 1–20, J.H. Steele (ed.), Plenum Press, New York.

Strickler, J.R. (1982). Calanoid copepods, feeding currents and the role of gravity. Science 218:158–160.

Strickler, J.R. (1984). Sticky water: a selective force in copepod evolution. In: Trophic Interactions within Aquatic Ecosystems, 187–239, D.G. Meyers and J.R. Strickler (ed.), Westview Press, Boulder, Colorado.

Vanderploeg, H.A. and G.A. Paffenhöfer (1985). Models of algal capture by the freshwater copepod Diaptomus sicilis and their relation to feed–size selection. Limnol. Oceanogr. 30–871–885.

Verity, P.G. (1985). Ammonia excretion rates of oceanic copepods and implications for estimates of primary production in the Sargasso Sea. Biol. Oceanogr. 3:249–283.

# CHAPTER 3

## When Should You Include Age Structure

by
Alan Hastings
Department of Mathematics and Division of Environmental Studies
University of California
Davis, CA 95616

## I.    INTRODUCTION

The theoretical ecologist faces a dilemma when deciding whether to include size and age structure in an attempt to understand species dynamics. There is a large cost in terms of difficulty, so the benefit in terms of increased understanding must be substantial. This question of whether to include age structure (or size structure) in models is really a question of scale — organizational scale. At the organizational scale of species interactions, when can one ignore information at a lower level, namely at the level of differences among individuals?

Many models of interacting species ignore the effects of differences among the individuals in each species. The same is true of the simplest one species models as well. Differences among individuals can be due to location in space, genetic differences, differences in age, size, or other state variables. A recent review by Werner and Gilliam (1984) stresses the importance of size structure in understanding the dynamics of populations. Moreover, recent advances in mathematical techniques (see for example the review in Metz and Diekmann, 1986) have made size and age structure easier to incorporate in some models. In this paper I attempt neither an exhaustive review of the biological evidence or of the modelling attempts or techniques used to treat age structure. Instead, I use several examples how age structure may affect construction of models in ecology, and the consequent implications of using age structure to explain ecological phenomena.

The structure of the current paper is to provide a number of examples where age structure is vital to understanding the dynamics of the populations considered, thus justifying its inclusion in models at any cost. The goal is to try to determine if there is a common theme to these examples, so the importance of including age structure can be determined *a priori*. A note of caution is that many authors, including Werner and Gilliam and Metz and Diekmann have argued that age rarely is the primary variable of interest, which may be size or developmental stage. However, in insects, there may often be a good correlation between age and development.

Before turning to the examples, I suggest that the basic rationale for ignoring age structure, at least in equilibrium questions, can be traced to the work of Lotka, who demonstrated that in a single species without density dependence, a stable age distribution is approached in general. I will now begin discussing a series of examples, first dealing with single species models of intraspecific competition, cannibalism, and dispersal. I will then turn to two species models and deal with examples of the two basic interactions, namely competition and predation.

## II. SINGLE SPECIES MODELS

The first example I will present is one given in Gurney, Nisbet, and Lawton (1984) concerning the dynamics of the moth *Plodia interpunctella*. Counts of dead moths showed dramatic oscillations in an experiment running 600 days, where the oscillatory period was on the order of 30 days. The important features of a caricature of the life cycle of this moth can be included in a model with only two life stages, adults and larvae. One important feature, characteristic of many arthropods, is the existence of distinct life stages, where all but the last are of fixed length. The question asked is whether competition among the larval stage can lead to cycles.

The model developed by Gurney, Nisbet, and Lawton was of the following general form. Let $l(t,a)$ be a density function on $a$ for the number of individuals. Let $\tau_1$ be the length of the larval stage. A general model for the dynamics of the larval stage is given by the von Foerster (or McKendrick) model:

$$\partial l/\partial t + \partial l/\partial a = -\mu_1 l, \quad 0 < a < \tau_1, \tag{1}$$

where $\mu$ is the per capita death rate of larvae. The birth rate, which depends on the total number of adults, $A$, is given as

$$l(0,t) = \beta A. \tag{2}$$

Letting $\alpha(t,a)$ be a density function on age $a$ at time $t$ for the number of adults, the adult dynamics is given by:

$$\partial \alpha/\partial t + \partial \alpha/\partial a = 0, \quad 0 < a < \tau_a, \tag{3}$$

where the rate of production of adults of age 0 equals the larvae of age $\tau_1$:

$$\alpha(t,0) = l(t,\tau_1). \tag{4}$$

The question of biological interest lies in the incorporation of larval competition in the specification of the function $\mu_1$. There are two extreme ways to incorporate this competition. One is to assume that the increase in the death rate of the larvae depends on the total number of larvae, so

$$= \gamma \int_0^{\tau_1} l(t,a) \, da, \tag{5}$$

where $\gamma$ is some constant. The other possibility is that only larvae of the same age compete with each other, so

$$\mu_1(t,a) = \gamma \, l(t,a). \tag{6}$$

The results obtained by Gurney, Nisbet, and Lawton show that the dynamic behavior of these two cases is very different. For the death rate (5), no sustained oscillations are possible, while for the case (6) sustained oscillations are possible.

The theme which emerges is that interactions among narrow age classes can be destabilizing, and lead to very different dynamic behavior. Further, the mere presence of larval competition is not enough, but that the manner in which the ages of the larvae interact is of prime importance. This affirms the importance of including age in models, even within a life stage such as larvae.

Similar cases of competition within a juvenile stage leading to oscillations are found in the general context of recruitment models for fisheries. One well studied example is a model for recruitment into striped bass populations studied by Levin and Goodyear (1980). Another example where this kind of interaction can lead to interesting dynamic and equilibrium behavior is in the models of Dungeness crab studies by Botsford (1981).

*Tribolium sp.*

Another case where the role of age structure becomes apparent is in cannibalism. In recent studies (Hastings, 1987, Hastings and Costantino, 1987) with R. F. Costantino, I have begun looking at models for cannibalism in *Tribolium* sp. These models incorporate a fixed time interval for the egg stage and the larval stage, which cannibalizes the egg stage. Using a separation of time scales approach, (i.e., we examine egg and larval dynamics, and ignore adults under the assumption that their dynamics occur on a slower time scale) we claim that the dynamics can be explained by focussing on these two stages alone. This case is actually very similar to the example of competition discussed above, since in this model there is no benefit accrued to the cannibalistic stage. One way competition would lead to the same model.

Once this assumption is made, we then proceed to analyze the egg–larval model, which is again phrased originally in terms of the McKendrick − von Foerster equation (see Diekmann et. al. 1986):

$$\partial n(t,a)/\partial t + \partial n(t,a)/\partial a = -\mu(t,a)n(t,a), \tag{7}$$

where $n(t,a)$ is the age distribution in the population at time $t$ and $\mu(t,a)$ is the time and age specific death rate. Cannibalism is included in this model in the form of the death rate $\mu(t,a)$, which in fact will be a functional of the population size at different ages.

Here, ignore changes in the size of the adult population (the fast time scale assumption), so the total birth rate is taken as a constant. The boundary condition becomes

$$n(t,0) = b. \tag{8}$$

Identify individuals with ages a between 0 and $A_e$ as eggs, and individuals with ages between $A_e$ and $A_e + A_1$ as larvae, and ignore any individuals older than this. Denote the number of larvae by $N_1$.

As in Hastings (1987) and Hastings and Costantino (1987), use a linear functional response, so the death rate of eggs from cannibalism by larvae will be $c_1 N_1(t)$. Let the death rate of eggs due to causes other than cannibalism by larvae be denoted by $\mu_e$ and the death rate of larvae by denoted by $\mu_1$. Thus the function

$$\mu(t,a) = \begin{cases} \mu_e + c_1 N_1(t), & 0 < a < A_e \\ \mu_1, & A_e < a < A_e + A_1. \end{cases} \tag{9}$$

Denoting the number of eggs by $N_e(t)$, this leads to (see Hastings, 1987): The following equation for $N_\ell(t)$:

$$N_1(t) = \int_0^{A_1} b \exp\{-\mu_e A_e - c_1 \int_0^{Ae} N_1(t-a-s)\,ds\}\exp(-\mu_1 a)\,da. \tag{10}$$

As in Hastings (1987) and Hastings and Costantino (1987), I will present results under the assumption that the death rate of larvae is so small that it can be taken as zero as a first approximation:

$$\mu_1 = 0. \tag{11}$$

A linear stability analysis of (9) in Hastings (1987) shows that the equilibrium of (10) is locally asymptotically stable if and only if:

$$bc_1 \exp(-\mu_e A_e) < \gamma^* \exp(\gamma^* A_1 A_e) \tag{12}$$

where:

$$\gamma = c_1 N_1 / A_1, \tag{13}$$

where $N_1$ is the equilibrium value determined from (10), and

$$\gamma^* = \omega^2 / [2 - \cos(A_e \omega) - \cos(A_1 \omega)] \tag{14}$$

and

$$\omega = 2\Pi / (A_e + A_1). \tag{15}$$

Moreover, stability is lost by a pair of complex eigenvalues crossing the imaginary axis with nonzero speed which implies that there is a Hopf bifurcation and small amplitude periodic orbits result (Diekmann and van Gils, 1984).

In Hastings (1987), the stability of the bifurcating solutions was determined using a formula of Diekmann and van Gils (1984). If $\mu_1 = 0$, the model (9) always undergoes a subcritical Hopf bifurcation at the point $\gamma = \gamma^*$, meaning that the bifurcating orbits were unstable in this case. The result carries over to the case where $\mu_1$ is sufficiently small: the model (9) still undergoes a subcritical Hopf bifurcation at the point $\gamma = f(\mu_1)$, where in the limit as $\mu_1$ approaches zero, $f(\mu_1)$ approaches $\gamma^*$. These results (plus the obvious boundedness of solutions to (10)) lead to the conclusion that for $\gamma^* - \gamma$ positive and sufficiently small, the model (10) has at least two different attractors, one a stable equilibrium and the other a periodic or more complex solution. Moreover, the length of the cycles can be shown by numerical work to equal to the sum of the larval period plus the egg period, which for *Tribolium* can be quite short, roughly 18 days. Thus, experiments which typically have gathered data at 14 or 10 day intervals may be taking too coarse a sample.

Thus, this example is very similar to the competition example previously discussed. The form of the interaction leads to cycles. Thus, one theme that emerges is that age structure can play an important role in the generation of population cycles. Both examples, however, show that the details of the interaction are important in determining whether cyclic behavior, in fact, occurs. In both examples, the details provided by age structure are both necessary to match, even crudely experimental data, and provide guides to planning experiments.

Age Structure and Dispersal

I will now turn to one more way in which age structure can affect the dynamics of a single species. In this case it is through interaction with dispersal (Hastings, in prep.) For purposes of illustration, assume that one can break up a species into two age classes, juvenile and adult, with equal lengths. Further make the assumption that there is strong density dependence in the juveniles, while the adults form the dispersing stage. This is at least a caricature of the life cycle of many arthropods. The model thus takes the rough form:

$$\begin{bmatrix} B_0(n_0,n_1) & B_1(n_0,n_1) \\ S(n_0,n_1) & 0 \end{bmatrix} \begin{bmatrix} n_0 \\ n_0' \end{bmatrix} = \begin{bmatrix} n_0' \\ n_1' \end{bmatrix}$$

(16)

where $n_0$ is the numbers in age class 0, $n_1$ is the numbers in age class 1, and $B_i$ are density dependent birth rates while S is a density dependent survival rate. The numbers in the following year are $n_0'$ and $n_1'$. Then the conditions for stability are that:

$$2 > 1 + ad - bc > |a+d| \tag{17}$$

where

$$a = B_0 + (\partial B_0/\partial n_0) n_0 + (\partial B_1/\partial n_0) n_1$$
$$b = B_1 + (\partial B_0/\partial n_1) n_0 + (\partial B_1/\partial n_1) n_1$$
$$c = S + (\partial S/\partial n_0) n_0 \tag{18}$$
$$d = (\partial S/\partial n_1) n_0$$

Now consider a two patch version, where before the population growth described by (16) a fraction $D_i$ of age class i leave each patch and settle randomly in the two patches. The stability conditions for this model are:

$$1 > (1-D_0)(1-D_1)(ad-bc) \tag{19}$$

$$1 + (1-D_0)(1-D_1)(ad-bc) > |(1-D_0) a + (1-D_1) d|. \tag{20}$$

First note that (17) implies (19) so that the interesting condition is (20). I will mention one special case of interest. Assume that all dispersal takes place in age class 1. Then there will be a range of values of $D_1$ for which (20) will fail if and only if

$$|a| > 1. \tag{21}$$

A similar condition holds if only age class 1 disperses, namely

$$|d| > 1. \tag{22}$$

The important point is that (17) can hold even if (21) or (22) does. Thus the conditions for stability of a spatially uniform equilibrium are more restricted than the conditions for stability in a single population. Moreover, the conditions that lead to this effect: strong density dependence in one age class and dispersal in the other, may be realistic.

This result shows that a combination of age and spatial structure can lead to a "Turing effect" (see Levin, 1974), where without dispersal there is a stable, spatially uniform equilibrium, while with dispersal, the uniform equilibrium becomes unstable. Differences among individuals within a single species must be included for this effect to be noticed. This example differs from the previous two because the effects of age structure only show up in the interaction with dispersal. This further illustrates the role played by the details of population structure in the dynamics of populations.

## III. TWO SPECIES MODELS

I will now discuss to two species models, again trying to determine when age structure is likely to be important for understanding the dynamics of interacting species. I will give one example of competition and one of predation. The discouraging aspect of these results is that the interaction of age structure with other factors appears more subtle than in single species models.

## Competition

The first example I will mention is one of competition between two bruchid beetles, *Callosobruchus chinensis* (*L.*) and *C. maculatus* (*F.*) studied by Bellows and Hassell. They studied the competitive interaction between these two beetles in the laboratory, where the outcome of competition is that *C. chinersis* always wins. They then proceed to construct a series of models to explain this outcome, in each case providing an estimate of parameters from their laboratory work. The first model they attempt to use is a single age class model with nonlinear growth rates. In this model there is an unstable equilibrium (outcome of competition dependent on starting conditions). In the two age class model they construct the outcome is the same. Only when they go to a 'systems' model, which is examined through computer simulation, does the model outcome match the experimental results. They suggest that the reason for this is that the superiority of *C. chinersis* in competition is due to its shorter development time. This is a feature which would only show up in detailed age–structured models.

Predation

I will now discuss an example of a class of predator prey systems I previously studied where the role of age structure can be very important. Moreover, the role of life stages of fixed length (as opposed to variable length) in determining the importance of age structure clearly emerges in these examples. This is a system of age dependent predation, as discussed in Hastings (1983). Assume that there are two distinct life stages in the prey: juvenile and adult. Assume that predation occurs only on the adult stage, with the fecundity of all adults the same. This leads to the following general system of equations:

$$dH/dt = r \int_0^\infty H(t-s)G(s)ds = DH - Pf(h)$$

$$dP/dt = cPf(H) - k\,P,$$

where $H(t)$ is the number of adult prey at time $t$, $P(t)$ is the number of predators, $G(z)$ is the probability that an individual survives to age $z$ and matures from juvenile to adult at age $z$, $r$ is the per capita birth rate, $D$ is the death rate of adult prey from causes other than predation, $k$ is the death rate of prey, $c$ is a conversion factor, and $f(H)$ is the functional response (Murdoch and Oaten, 1975).

The model will be completely specified once the functional response, $f(H)$, and the function $G(z)$ are given. In what follows only derivatives of $f$ at equilibrium are used (cf. Murdoch and Oaten, 1975). Results then depend on the form of $G(z)$. I assume the functional response is destabilizing. With a fixed juvenile period, there is a wide range of the parameters over which the system can cycle, and the local stability of the equilibrium depends critically on the relationship among the parameters. Moreover, increasing delays (lengths of the juvenile stage) can either be stabilizing or destabilizing. In contrast, if the probability of maturation is independent of the age of the juvenile, then the system is much more stable. Also, there is a critical mean length of the juvenile period below which the system is (locally) unstable, and above which it is stable.

The general themes of complicated behavior and critical dependence on the parameter values, as well as the fact that systems with fixed stages are more complicated, have emerged in many other studies as well. This work is extensively reviewed in the book by Metz and

Diekmann (1986).

## IV. CONCLUSIONS.

The overall theme of the examples I have discussed here is that age (or size) structure is important for understanding the dynamics of single and interacting species. Moreover, the details of the age structure and age dependence of the interactions will be important in general. If this was the only conclusion, it would suggest that theoretical ecology would at best become a series of special cases. Some general principles do seem to emerge, both at the level of building blocks for models and at the level of outcomes of models.

First, there are some general guides as to when age structure is most likely to have an important impact on the dynamics of single or interacting species. One rule of thumb is that discrete life stages of fixed length are more likely to lead to complicated and nonintuitive results than are life stages of variable length. This echoes a theme suggested by Chesson both in this volume and elsewhere about the stabilizing effect of variability. (Given the appearance of the same principle at different levels of organization, this is also an example of how the building blocks of models may remain the same, although the 'buildings' we construct may differ from system to system.)

A second rule of thumb is that not only are distinct life stages important in generating important effects of age structure, but so is making the different biological interactions dependent on the ages. This is illustrated by the example of Plodia interpenctella.

The third conclusion is more pessimistic. As the two examples chosen above show, typically it becomes more difficult to construct and analyze age structured models as the number of species is increased, even to two (see also further examples in Metz and Diekmann). However, in both of the cases cited above the role of age structure is very important, and its effects can become more subtle, depending on particular details of the age structured interaction. I would suggest that given the current state of knowledge, the role of age structure in multispecies interactions may be very important, but much more study is needed. Perhaps, returning to the theme of this volume, if the questions asked about more complex systems require less detail, one may be able to ascertain when age structure can be ignored.

## REFERENCES

Bellows, T. S. Jr. and M. P. Hassell, 1984 Models for interspecific competition in laboratory populations of *Callosobruchus* spp. J. Anim. Ecol., 53: 831–848.

Botsford, L. W., 1981 The effects of increased individual growth rates on depressed population size. Amer. Nat., 117: 38–63.

Diekmann, O., R. M. Nisbet, W. S. C. Gurney, and F. van den Bosche., 1986 Simple mathematical models for cannibalism: a critique and a new approach. Math. Biosci., 78: 21–46.

Diekmann, O. and Gils, S. A. van, 1984 Invariant manifolds for Volterra integral equations of convolution type. J. Diff. Equ., 54: 139–180.

Gurney, W. S. C., Nisbet, R. M. and Lawton, J. H., 1983 The systematic formulation of tractable single species population models incorporating age structure. J. Anim. Ecol., 52: 479–485.

Hastings, A., 1983 Age dependent predation is not a simple process. I. Continuous time models. Theor. Pop. Biol., 23: 347–362.

Hastings, A. and R. F. Costantino, 1987 Cannibalistic egg–larval interactions in *Tribolium*: an explanation for the oscillations in population numbers. Amer. Nat., 130: 36–52.

Hastings, A., 1987 Cycles in cannibalistic egg–larval interactions. J. Math. Biol., 24: 651–666.

Levin, S. A. and Goodyear, C. P., 1980 Analysis of an age structured fishery model. J. Math. Biol., 9: 245–274.

Metz, J. A. J. and Diekmann, O., 1986 *The Dynamics of Physiologically Structured Populations*. Springer–Verlag, New York.

Werner, E. E. and Gilliam, James F., 1984 The ontogenetic niche and species interactions in size–structured populations. Ann. Rev. Ecol. Syst., 15: 393–425.

CHAPTER 4

Spatial Aspects of Species Interactions:
the Wedding of Models and Experiments

P. Kareiva
M. Andersen
Department of Zoology
University of Washington
Seattle, WA 98195

## INTRODUCTION

Ecological theory is in an obvious state of flux. Earlier attempts to model communities using equilibrium analyses of Lotka–Volterra equations have fallen into disfavor (see the volumes edited by Price et al. 1984, and by Diamond and Case 1986). Indeed, it seems that many experimental ecologists have given up on theory altogether — rather than testing models, field ecologists are now mostly concerned with proper experimental design and the detection of significant effects attributable to one manipulation or another. Theoreticians, of course, continue to develop models; but even among theoreticians there is no consensus about the sorts of models that most deserve our attention. While everyone may agree that we need to move beyond simple Lotka–Volterra equations, few can agree upon the ways in which models might be made more pertinent to the natural world. For example, in this volume alone, arguments are made for extending theory to include: age structure (Hastings), stochastic environments (Chesson), the effects of spatial scale (Levin), fluid dynamics (Okubo), and food web architecture (Cohen, Pimm, and Yodzis). Yet clearly no model can handle all of these complications at once. This leaves us with the problem of determining which theoretical elaborations are called for by different real–world systems, and how these theories might be applied to experimental studies.

Our approach emphasizes models of <u>dispersal and spatial structure</u> as the theoretical elaboration we find especially useful. Because we are disturbed by the independence of theory and empiricism in contemporary ecology, we focus on applications of models to field studies rather than on the analyses of the models themselves.

We begin with a brief overview of the importance of dispersal and spatial heterogeneity as suggested by previous work. This leads us to point out some limitations of conventional species–removal or species–addition experiments in those circumstances where dispersal and heterogeneity markedly influence population dynamics. We argue that experimental studies of interacting species can be made more informative if combined with mathematical models that detail how organisms move about in space. Two specific situations command our attention: (i) the interaction between ladybird beetle predators and their patchily distributed aphid prey, and (ii) the invasion of open habitat by perennial plants. Finally, using these specific investigations as a jumping–off point, we conclude with some general remarks about future directions for theoretical ecology.

## II. SPATIAL FACTORS IN THEORY AND PRACTICE

Mathematical ecologists have produced numerous models showing how spatial aspects of species interactions can provide opportunities for coexistence (Levin 1974; Hastings 1977, 1978) and for mosaic patterns (Segel and Jackson 1972, Mimura 1984). Without the benefit of equations, naturalists have similarly concluded that many species persist together only because of the opportunities provided by spatial heterogeneity and dispersal (see for example, Elton 1949). The key idea in both this mathematical and verbal theory is that if organisms are allowed to move around in a sufficiently large or sufficiently heterogeneous area, then multispecies patterns of coexistence may emerge which could not occur in smaller or simpler areas.

The role of the spatial dimension in species interactions is much more than an issue in theoretical ecology — it is a practical concern for anyone designing experimental studies of natural communities. For example, suppose a researcher wants to know whether a particular predator can hold a prey species at low densities, or whether one species competitively excludes another. The answers obtained from observations in small cages or small quadrats could be quite different from the dynamics that occur when organisms are allowed to range freely over large areas. After all, the resounding message of theory is that some dynamics, and some patterns of coexistence become possible only if species are allowed to interact in sufficiently large arenas. Natural communities occupy acres of land, not the few square meters that typify experimental manipulations of communities. To quantify just how real is the problem of spatial scale in experimental ecology, we searched Ecology from January 1980 to August 1986. We identified every paper in which the authors experimentally manipulated resources or populations in order to learn about community dynamics or community properties. We excluded studies of interacting species in which the focus was the individual, such as investigations of herbivore impact on plant reproduction or plant fitness. We also excluded studies of microorganisms. Our intent was to focus on the classical domain of community ecology — how interactions among species shape community structure and patterns of coexistence. We found 97 papers that satisfied our criteria of experimental community ecology. For each of these papers we recorded the maximum linear dimension of any experimental plot used in the research, and the maximum number of replicates of any treatment. (Thus, if a study included a 5m x 5m treatment replicated four times, and a 20m x 5m control replicated twice, we would record 4 as the replicate number and 20m as the linear dimension.) The results of this survey are summarized in figure 1. Viewed pessimistically, figure 1 suggests that ecologists tend to study nature by collecting data from but a few tiny plots. However, well–designed may be the experiments represented in figure 1, there is cause for discomfort. Over 45% of the papers we looked at included at least one treatment that was replicated no more than twice. Nearly one–quarter of the studies used plots no larger than .25m in diameter; one–half of the studies used plots no larger than a meter in diameter. One has to wonder whether studies conducted at such a small scale are not missing key aspects of species interactions.

Figure 1. Size and replication in experimental community ecology. Each point
    is taken from a different published paper in <u>Ecology</u> between January 1980 and August
    1986. See text for details.

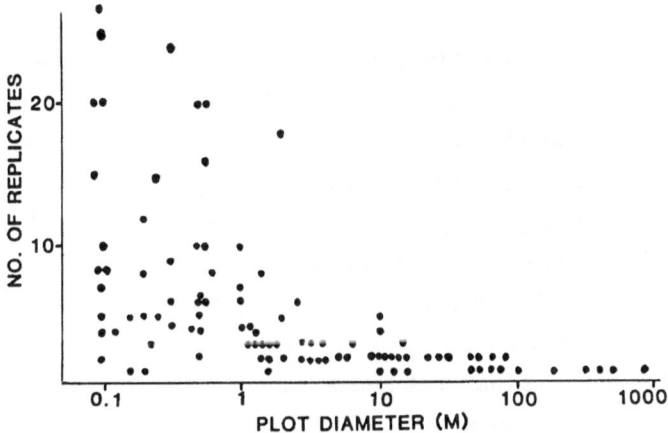

We do not mean to unfairly criticize the replication and size of experiments in community ecology. Several of our own studies fall in embarrassing positions on figure 1. The point is that replication and scale require money and personpower beyond the means of most ecologists. This is where theory becomes valuable. With exhaustive empiricism impractical, theory can direct experimental efforts toward promising questions. Theory is particularly useful where spatial scale is thought to be important to species interactions. For example, small–scale studies or observations can be used to develop specific models of interacting and redistributing populations. These models may generate predictions for large–scale dynamics, which can then be tested with a few judiciously planned large–scale experiments. This is the approach we try to follow in much of our own research: microscale observations are used to build models that address dynamics at the scale of tens or hundreds of meters. With a model firmly in mind, a single experiment can provide a powerful test of one's understanding of spatially–distributed interactions.

## III.   SPATIALLY DISTRIBUTED PREDATOR–PREY INTERACTIONS

    For over a decade theoreticians have been modelling the influences of dispersal and spatial patchiness on predator–prey systems. Much of this work has been inspired by

Huffaker's (Huffaker 1958) famous laboratory experiments using predator mites and prey mites. Huffaker's widely quoted conclusion was that predator–prey interactions can be stabilized by habitat heterogeneity, a theme that has been corroborated by several models (e.g., Hastings, 1977, Hilborn 1975, Maynard Smith 1974). However, most of these models are so abstract and general that they are difficult to relate to any real predator–prey interaction — they are what might be called "metaphors for nature". Nonetheless, the models serve the valuable purpose of suggesting field studies of habitat patchiness as worthy investments of time and labor.

Motivated largely by models such as Hastings (1977) and Maynard Smith (1974), we have been manipulating habitat patchiness in goldenrod fields so that we might determine the effect of patchiness on species interactions (Kareiva 1984). One experiment that has been replicated in three different fields and followed for four years involves mowing goldenrod fields into either a long continuous strip of vegetation (continuous treatment) or into a long archipelago of discrete patches of vegetation (patchy treatment). This experiment has revealed a striking effect of patchiness on a predator–prey system associated with goldenrod — the interaction between the ladybug Coccinella septempunctata and its aphid prey, Uroleucon nigrotuberculatum (for details on experimental design see Kareiva 1985). In twelve out of twelve cases (3 replicates x 4 field seasons) the Uroleucon aphids erupted into local outbreaks within the patchy vegetation but were held at bay by ladybugs in the continuous vegetation (Kareiva, 1987). This indicates that "patchiness" cannot comfortably be called stabilizing in this particular predator–prey system.

To understand why aphids outbreak in patchy goldenrod but not continuous goldenrod we decided to look carefully at the foraging behavior of Coccinella. Direct behavioral observations of Coccinella indicate that these ladybugs aggregate at clusters of aphids; if they aggregate rapidly enough they can suppress aphid population growth (Kareiva 1985). Patchiness leads to aphid outbreaks because it interferes with the effectiveness with which Coccinella aggregate at incipient aphid outbreaks.

The importance of predator aggregation to predator–prey dynamics has been a major theme of several models of arthropod predator–prey systems (Hassell 1978, Hassel and May 1974, 1985). But there is considerable uncertainty about whether theory or observations indicate that predator aggregation is stabilizing or destabilizing (contrast Hassell and May 1985 with Chesson and Murdoch 1986). When we viewed this uncertainty in the context of our Coccinella observations, it made sense to us — the consequences of aggregation are unpredictable because they depend on the details of prey population growth and predator foraging behavior. In order to capture this complexity, we decided to develop a model of a predator–prey system in which the process of aggregation was related to observations of individually foraging ladybugs. Our goal was to relate predator behavior to rates of aggregation and ultimately to the effectiveness of predators as regulators of prey populations.

We base our model on "area restricted search," a behavior found in numerous predators, including coccinellids (Banks 1957, Chandler 1969, Fleschner 1950, Nakamura 1984). Predators engaging in area restricted search increase their turning frequency immediately after finding and eating a victim. By specifying velocities and turning frequencies as a function of Coccinella gut fullness and hence prey density, we were able to derive the following expression for predator flux along the line x (see Kareiva and Odell, 1987 for details):

$$J_p = - D_p[V(x)] \frac{\partial P}{\partial x} + P X[V(x)] \frac{\partial V}{\partial x} \qquad (1)$$

where P is predator density, V is prey density, and Dp[·] and X[·] are functions of V. Because Dp and X vary with V, equation (1) describes a predator whose rate and direction of movement responds to spatial variation in prey density (which enters the equation through V(x) or $\partial V / \partial x$). In our particular derivation of (1), we explicitly relate Dp and X to ladybug velocity, turning frequency, and consumption rate. Dp turns out to be a random motility coefficient that increases with ladybug velocity and decreases with increasing aphid density; X is an aggregation coefficient in the sense that the larger X is, the faster ladybug populations will tend to move up gradients in prey density and thereby cluster at local aphid outbreaks. For our conceptualization of area restricted search, X increases with the square of ladybug velocity and has its greatest value where there is the largest change in ladybug turning frequency per unit change in prey density. Somewhat independent of the details, equations such as (1) must apply to most foraging predators since almost all predators adjust their movements to local prey abundance. The important thing is that Dp and X can be estimated from observations of individually foraging ladybugs; the observations needed for this task are shown in figure 2. Once Dp and X have been obtained, it is straightforward to extend equation (1) to the following spatially–varying predator–prey model:

$$\frac{\partial V}{\partial t} = Dv \frac{\partial^2 V}{\partial x^2} + F(V) - \phi(V,P) \qquad (2a)$$

$$\frac{\partial P}{\partial t} = \frac{\partial}{\partial x} (Dp[V(x)] \frac{\partial P}{\partial x} - PX[V(x)] \frac{\partial V}{\partial x}) \qquad (2b)$$

$$+ g(V,P)$$

where Dv is the prey's diffusion coefficient, F(v) represents prey population growth in the absence of predators, $\phi$ is the rate at which predators kill prey, and g is the source/sink term for local predator populations. By quantifying Uroleucon demography and Coccinella foraging, we have been able to completely specify all parameters and functions in equations (2a) and (2b) (Kareiva and Odell, 1987). To experimentally test the resultant model we created two different initial profiles of aphid and ladybug densities along a 10 meter goldenrod hedge (the leftmost graphs in figure 3). We then contrasted the spatiotemporal changes in aphid and ladybug densities predicted by the model to the observed spatio temporal dynamics (figure 3). The model is moderately successful, with most of the error stemming, we think, from an inaccurate forecast of aphid population growth (because we do not include age structure). But we do not see the main use of models such as equations (2a) and (2b) to be forecasting. Rather we hope to use the model to explore theoretical consequences of aggregation and to relate readily observed aspects of predator foraging (e.g. velocity and turning frequency) to the effectiveness with which predators aggregate at and contain incipient prey outbreaks.

Numerical analyses of the behavior of equations (2a) and (2b) offers some insight regarding the effects of predator aggregation on the dynamics of predator–prey systems (see also, Hassell and May 1985; Chesson and Murdoch 1986). We find that whether or not aggregating predators can contain an incipient prey outbreak depends quantitatively on the rate of aggregation relative to the rate of prey population growth. We do not need to speak phenomenologically about aggregation — we can make predictions about prey control on the basis of feeding rates, velocities of movement, and turning frequencies. For example, the

Figure 2. Behavioral input to aggregation model. In order to determine
    equation (1), curves such as those above must be obtained by observing individual
    ladybird beetles over short periods of time. Beetles need to be starved differing lengths
    of time for some suites of observations because starvation enters the model through its
    influence on gut fullness (see Kareiva and Odell 1987).

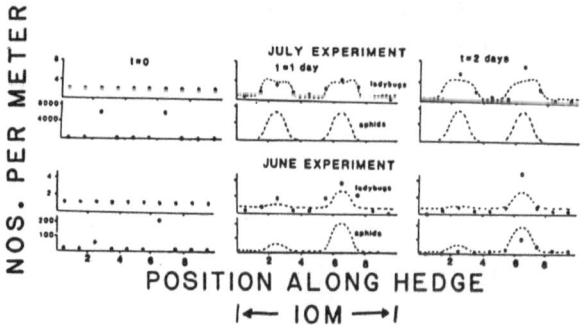

details of area restricted search behavior and the size of a prey outbreak interact to determine the rate at which predators aggregate, and as a result different predator species might be seen as being predisposed to taking advantage of (aggregating at) prey eruptions of different sizes. Numerical solutions of equation (2a) and (2b) also indicate that although a predator may aggregate at prey, this does not mean that mortality due to predation necessarily increases with increasing prey density. In particular, prey population growth may swamp predator aggregation and lead to predation rates which actually decline over some regions of increasing prey density. Thus it is unlikely that we can make general statements about what aggregation does to predator–prey systems. However satisfying they may seem, such simple theoretical predictions are unlikely to be valid for the panoply of behaviors that might be called predator aggregation. Nonetheless, a "conditional" theory that relates the influence of aggregation to the quantitative values of parameters such as predator mobility is within reach (see Kareiva and Odell 1987).

One other interesting feature of equations (2a) and (2b) is their ability to exaggerate the patchiness of prey distributions. In particular, when $\phi$, F, and g satisfy certain biologically plausible conditions (see Kareiva and Odell, 1987), the predator–prey system described by (2a) and (2b) can generate regular spatial patterns. Usually pattern formation models involve an initial steady state that is spatially homogeneous, but which, when perturbed, gives rise to stable spatial patterns (see Meinhardt 1982). In contrast, the mechanism for initiating pattern in (2a) and (2b) is <u>not</u> linear instability of a spatially homogeneous initial state. Instead, we think of pattern arising in <u>Coccinella</u> – <u>Uroleucon</u>

Figure 3. Testing the aggregation model using field experiments. Initial
profiles of aphid and beetle densities were arranged along 10 m goldenrod hedges and then observed for two days. The dashed lines represent the solution to equations (2a) and (2b) whereas the points are the observed mean ladybug and mean aphid densities (only in June experiment); the test is noncircular because the paramparameters for the model were estimated independently from the experiments shown above.

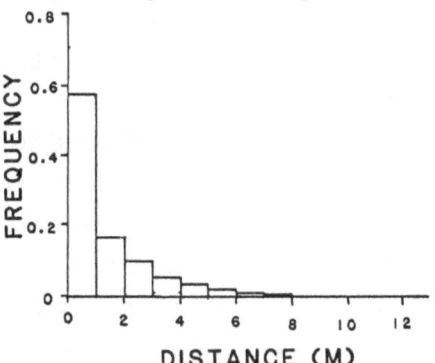

systems when a small aphid settlement escapes ladybug detection for long enough to grow to a very dense aphid population. Where and when this happens is a matter of chance. But once a local aphid outbreak has appeared, it will attract ladybugs. Some of those aggregating ladybugs will regularly wander away from the aphid peak [because of the diffusion term in (2b)] and suppress secondary aphid outbreaks in the vicinity of the original outbreak. The result can be a small peak of aphid density with a halo of ladybugs, isolated in the middle of vegetation largely free of aphids.

On the practical front we have been using equation (1) to organize a comparative analysis of several different ladybug species, each of which has some history of use in biocontrol programs. The ideal biocontrol agent should aggregate rapidly at incipient aphid outbreaks and then consume the aphids voraciously enough that they are locally annihilated. Using observations of ladybug searching behavior we have been attempting to predict the effectiveness with which different ladybug species find and contain patches of pest aphids in commercial pea and bean fields. Initially we were discouraged to find that Coccinella septempunctata, which readily aggregated at patches of goldenrod aphids (figure 3) did not do so when presented with aphids on field peas, and only slowly aggregated at clusters of aphids on row beans. In addition, the ladybug Hippodamia convergens aggregates dreadfully slowly, or not all, when presented with incipient outbreaks of aphids on either peas or beans. This led us to wonder whether equation (1) was appropriate only for the Coccinella – Uroleucon interaction. Inspection of the behavioral data has revealed that such is not the case. It appears that Coccinella's failure to aggregate when placed on peas and Hippodamia's generally bad performance makes sense in light of our derivation of equation (1). Recall that the rate of aggregation [i.e. $X$ in equations (1) and (2b)] is proportional to mean linear velocity and the amount by which ladybug turning frequency is reduced in regions of high aphid density relative to regions of low aphid density. Some plants (e.g., peas) are so architecturally complex that ladybugs foraging on these plants never attain rapid speeds and reverse directions so often in the absence of prey that there is little room for an increase in turning frequency, where prey are superabundant (see Figure 4). We think that plant architecture places a major constraint on the effectiveness of predators as biocontrol agents because it can dampen the aggregation response (see figure 4). Instead of outright rejecting equation (1), we now use it to characterize the interaction of ladybug species and plant architecture.

IV.    PLANTS IN SPACE: THE PROBLEM OF INVADING POPULATIONS

The spatial dimension also plays an obvious role in the dynamics of plant communities. For example, by growing on a particular site, a plant prevents other plants from surviving at, and possibly around, that location. It is no accident that many advances in the statistical modelling of spatial pattern have been motivated by examples involving plant populations (Cottam and Curtis 1949, Clark and Evans 1954, Moore 1954, Thompson 1956, Besag and Gleaves 1973). But although there are numerous statistical models describing the spatial patterning of plants, there are only a handful of models that describe how plants read and occupy space, much less how this process influences plant population dynamics. Ultimately, ecological theory needs to address the dynamics and interactions of dispersing, space–occupying plant populations. Schaffer and Leigh (1976) made this point over a decade ago, but concluded that mathematical models relevant to plant populations would be

Figure 4. How plant architecture might constrain predator aggregation. Any
feature of plant architecture that shifted the solid curves towards their dashed
counterparts would in turn reduce the rate at which predators could aggregate in
regions of high prey density.

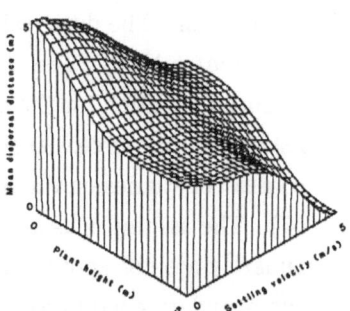

intractable. Recently theorists have finally begun to make progress with these problems. For
example, Pacala and Silander (1985) have developed a general theory for annual populations
by assuming a poisson dispersion of plants; but because this poisson assumption removes any
need of keeping track of spatial coordinates, their approach glosses over some aspects of
spatial dynamics we think are important. At a more abstract level, analyses of
continuous–space, discrete–time models with diffusion operators have revealed a wide range
of behavior, including chaos and non–uniform steady states (J.D. Murray, pers. comm.); but
these continuum models are not easily linked to known plant life cycles (they will, however,
be useful touchstones for more specific models).

In our studies of dispersing plant populations we have decided to focus on a naturally
simplified system, namely the barren areas created by the 1980 eruption of Mt. St. Helens.
This system is simplified in three ways (see also del Moral 1983; del Moral and Wood 1986):

1)      The devastation caused by the eruption has reduced habitat heterogeneity.
2)      The number of plant species in these areas is greatly reduced relative to similar
        habitats elsewhere in the Cascade Mountains. Thus, interspecific interactions may be
        assumed to be infrequent and unimportant.
3)      There are distinct borders between vegetated areas and devastated areas.


The plant we have been focusing on is Aster ledophyllus, a dominant wind–dispersed
reinvading perennial which is common in subalpine habitats throughout the Cascades.

The first step in modelling an invading plant population is to model the dispersal
process. Studies of wind–dispersal in plants have produced large quantities of data (see
especially McEvoy and Cox, 1987) and some good verbal theory (Sheldon and Burrows 1973,
Ellner and Shmida 1981, Venable and Levin 1983, Verkaar et al. 1983). Mathematical
analyses, however, have focused more on the aerodynamics of individual seeds than on
population–level consequences of dispersal (Burrows 1973, 1983). Therefore, there has been

practically no mathematical framework for interpreting seed shadow data in terms of population spread.

Following a reductionist program, we are attempting to predict seed shadows of single plants by considering the mechanisms involved in the transport of seeds by wind, and then working up to the level of an entire population. The dispersal of seeds by wind is a turbulent transport phenomenon (Okubo 1980, Schrodter 1960). The classical approach to these phenomena is to derive a partial differential equation for the concentration of suspended particles as a function of time and distance from source. Analytic solutions to such equations have been published, mostly in the context of studies on the dispersion of atmospheric pollutants (Sutton 1953, Godson 1957). The assumptions about boundary behavior that allow these solutions to be obtained unfortunately are not appropriate to the release of seeds from plants. Furthermore, when we assume boundary conditions appropriate to a plant releasing seeds, we end up with a problem apparently lacking analytical solutions.

In lieu of turbulent diffusion equations we have adopted an alternative approach to modeling Aster seed dispersal. The turbulent transport of seeds is physically equivalent to a Brownian motion process (with drift due to gravitational settling of the seeds) in the vertical (z) dimension plus convective transport by wind acting in the horizontal (x) direction. Thus, one can obtain an approximate seed shadow by simulating the trajectories of a large number of individual seeds. With this approach, one can account for the fact that wind speeds are not constant, but vary from time to time. Wind speed data from a weather station at our field site on Mt. St. Helens are approximately exponentially distributed with a mean velocity of 1.07 m/s during August and September (Reynolds and Bliss 1988); this is the time of the year during which the seeds of Aster ledophyllus are dispersing.

As far as the plant is concerned, the crucial parameters for seed dispersal are the release height of the seeds and the settling velocity of the seeds. If there is such a thing as an optimal dispersal distance for a plant, that optimum must be achieved by evolutionary adjustment of these two parameters. Mean dispersal distances from simulations using different values of settling velocity and plant height are shown in figure 5 (the release height and settling velocity for Aster seeds is given by a star). To obtain this figure we simulated the trajectories of 1000 seeds at each of the indicated points in the parameter space; wind velocities for each seed in each simulation were drawn from an exponential distribution. Note that settling velocity influences dispersal distance more markedly than does release height. Quantitative results such as this allow us to infer the consequences of genetic variation in seed morphology. Using these results we are determining whether
there are any particular colonizing cohorts of plants that represent superior dispersers with respect to their position in the morphological space defined by figure 5.

Now that we are able to predict the seed shadow of an individual plant we are also in a position to model reproduction and dispersal in empty habitats. For an annual plant with a net reproductive rate of $\alpha$, and seed shadow $v(x)$, the population size in the nth generation at position $x$ is

$$P_n(x) = \alpha \int_{-\infty}^{+\infty} P_{n-1}(\delta)\, v(x - \delta)\, d\delta \qquad (3)$$

Here $v(x)$ has been rescaled so that $\int_{-\infty}^{+\infty} v(x)\, dx = 1$, i.e., so that $v(x)$ is a probability density function. Results of Weinberger (1978) can be applied to this model to show that the plant population will invade an area at an asymptotic rate that depends on $v(x)$. The two

Figure 5. Predicting dispersal distance from the release height of seeds and
their settling velocity. Different settling velocities correspond to the different
series of connected points — the number to the right of the topmost point gives the
settling velocity (in m/sec) for that point and its connected points. We explored
the above parameters because they surround the position of Aster in the parameter
space (indicated by the star).

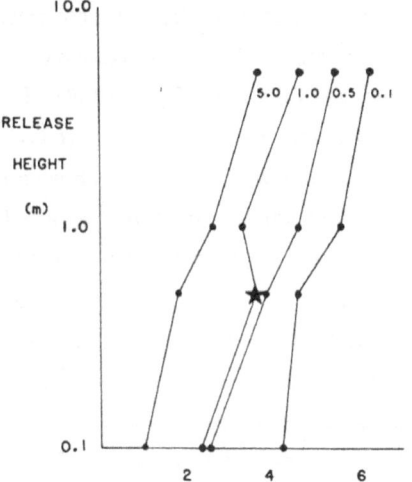

functions most commonly used to describe $v(x)$ are a normal distribution centered at $x = 0$ and a double exponential distribution centered at $x = 0$ (this is the distribution predicted by the seed dispersal simulation described earlier and also typically observed in field studies). If $v(x)$ is a normal density function with standard deviation $\sigma$, the mean dispersal distance is $0.675\sigma$. If the distribution of dispersal distances is double exponential, parametrized so that the standard deviation is again given by $\sigma$, the mean dispersal distance is $0.490\sigma$. The asymptotic propagation velocity is higher if $v(x)$ is double exponential than if $v(x)$ is normal (Mollison 1977). Models such as this can be applied to field populations of annual plants and are clearly better suited to the synchronous life–cycles of plants than continuous–time models such as those of Skellam (1951).

Unfortunately, Aster ledophyllus is not an annual, and annuals are totally absent from the recolonizing subalpine plant community at Mt. St. Helens. Thus, we require a more complex model than that of equation (3). For a perennial plant with two age classes (immatures and reproductives), we can generalize equation (3) to the system

$$P_{1,t}(x) = \left[\alpha \int P_{2,t-1}(\text{\v{s}}) \, v(x - \text{\v{s}}) d\text{\v{s}}\right] + \left[\beta P_{1,t-1}(x)\right]$$

$$- \left[\beta^{m-1}\alpha \int P_{2,t-m}(\text{\v{s}}) \, v(x - \text{\v{s}}) \, d\text{\v{s}}\right]$$

$$P_{2,t}(x) = \left[\beta^{m-1}\alpha \int P_{2,t-m}(\text{\v{s}}) \, v(x - \text{\v{s}}) \, d\text{\v{s}}\right] + \gamma P_{2,t-1}(x)$$

Here $P_{1,t}(x)$ and $P_{2,t}(x)$ are the populations sizes of immatures and reproductives, respectively, at time t. Age at first reproduction is m; $\beta$ and $\gamma$ are immature and adult survivorships, respectively, and $v(x)$ and $\alpha$ are as in equation (3).

Analytical results for (4) are hard to come by. We conjecture that solutions will show travelling–wave behavior with a minimum asymptotic velocity as is typical for interaction–redistribution systems (see Levin and Segel 1986). It would be interesting to know how far behind the advancing front one has to go before coming to a stable age distribution. Numerical studies are under way to explore these questions.

In any event, no advancing front has yet appeared in populations of Aster ledophyllus recolonizing Mt. St. Helens. The evident reason for our model's failure is its neglect of interactions between plant species. Aster, although it is a fairly good disperser, appears to require a better–than–average microenvironment to survive as a seedling. In the field, such microsites are typically provided by "nurse plants" (adults of other species, which trap dispersing seeds and shelter dispersing seedlings). Thus, it seems that our "naturally simplified system" was not as simple as we had thought; we find this outcome distressing, but hardly surprising. The logical next step is to incorporate nurse plant effects into the model and see whether this added realism makes the model's predictions correspond more closely to field data. Again, we do not intend the model to mirror reality, but to interact with our experimental studies so that theory and data can inform each other.

## V.  CONCLUSIONS

The models we have described are in no way general — they apply only to the particular experiments or situations for which they were developed. It is hard for us to imagine that any general spatially–distributed model could effectively describe a real system. The general theory is necessary, however, to guide the development of specific models. For instance, our aggregation model drew its inspiration from earlier equations due to Keller and Segel (1971) and Hassell and May (1974). Our seed dispersal models are extensions of simpler diffusion equations discussed by Okubu (1980). Furthermore the initial motivation for carving goldenrod into strips and archipelagos was provided by the purely theoretical papers of Hastings (1977, 1978) and Caswell (1978); similarly our focus on predator aggregation was sparked by Hassell and May's (1974, 1985) "aggregation" models. It is silly and unwise to demand that general models be directly testable — they should be sources of ideas, not pencil and paper replicas of nature.

But we do think theoretical ecology has suffered from a lack of specific models that are tailored to experimental systems. It is only through specific models that we can learn under which circumstances particular branches of general theory will be most helpful. In addition, when it comes to systems in which dispersal or spatial heterogeneity are thought to be important, a purely experimental approach is inadequate. Without a model, how can the fact that a plant disperses its seed an average distance of one meter be used to predict the rate at which a population will invade open habitats? Without models, how can we relate differences in ladybug foraging behavior to quantitative assessments of species as biocontrol agents? Because intuition is such a poor guide to the dynamics of interaction and redistribution systems, studies of spatially–distributed species interactions require the wedding of experiments and theory — neither approach by itself will get us very far.

## ACKNOWLEDGEMENTS

This research was supported by NSF grants BSR 8517183 and BSR 8605303 to P. Kareiva and grant GR/C/63595 from the Science and Engineering Council of Great Britain to the Center for Mathematical Biology at Oxford University. Many of the results and ideas in this paper are due to Garrett Odell. We thank N. Cappuccino and A. Power for comments on an earlier version of the manuscript.

## REFERENCES

Banks, C. 1957. The behavior of individual Coccinellid larvae on plants. Brit. J. Anim. Behaviour 5: 12–24.

Besag, J. and J.T. Gleaves. 1973. On the detection of spatial pattern in plant communities. Bulletin of the International Statistical Institute 45: 153–158.

Burrows, F.M. 1973. Calculation of the primary trajectories of plumed seeds in steady winds with variable convection. New Phytologist 72: 647–664.

Burrows, F.M. 1983. Calculation of the primary trajectories of seeds and other particles in strong winds. Proceedings of the Royal Society of London, series A 389: 15–66.

Chandler, A. 1969. Locomotory behavior of first instar aphidophagous Syrphidae (Diptera) after contact with aphids. Anim. Behav. 17: 673–678.

Chesson, P. and W. Murdoch. 1986. Aggregation of risk: relationships among host–parasitoid models. Am. Nat. 127: 696–715.

Clark, P.J. and F.C. Evans. 1954. Distance to nearest neighbor as a measure of spatial relationships in populations. Ecology 35: 23–30.

Cottam, G. and J.T. Curtis. 1949. A method for making rapid surveys of woodlands, by means of pairs of randomly selected trees. Ecology 30: 101–104.

del Moral, R. 1983. Initial recovery of subalpine vegetation on Mount St. Helens, Washington. American Midland Naturalist 109: 72–80.

del Moral, R. and D.M. Wood. 1986. Subalpine vegetation recovery five years after the Mount St. Helens eruptions. In S.A.C. Keller (ed.) Mount St. Helens: Five Years Later. Symposium proceedings, Eastern Washington University, Cheney, Washington. In press.

Diamond, J. and T. Case. 1986. Community Ecology. Harper and Row, New York.

Ellner, S. and A. Shmida. 1981. Why are adaptations for long–range seed dispersal rare in desert plants? Oecologia 51: 133–144.

Elton, C. 1949. Population interspersion: an essay on animal community patterns. J. Ecology 37: 1–23.

Fleschner, C. 1950. Studies on the searching capacity of the larvae of three predators of the citrus red mite. Hilgardia 20: 233–265.

Godson, W.L. 1957. The diffusion of particulate matter from an elevated source. Archiv. fur Meteorologia, Geophysik, und Bioklimatologie A. 10: 305–327.

Hassell, M. 1978. The dynamics of arthropod predator–prey systems. Princeton University Press, Princeton, New Jersey.

Hassell, M. and R. May. 1974. Aggregation in predators and insect parasites and its effect on stability. J. Anim. Ecol. 43: 567–594. J. Anim. Ecol. 43: 567–594.

Hassell, M. and R. May. 1985. From individual behavior to population dynamics, in Behavioral Ecology, R. Sibley and R. Smith, editors, British Ecological Symposium, Blackwell, Oxford, England.

Hastings, A. 1977. Spatial heterogeneity and the stability of predator–prey systems. Theor. Pop. Biol. 12: 37–48.

Hastings, A. 1978. Spatial heterogeneity and the stability of predator–prey systems: predator–mediated coexistence. Theor. Pop. Biol. 14: 380–395.

Kareiva, P. 1984. Predator–prey dynamics in spatially–structured populations: manipulating dispersal in a coccinellid–aphid interaction, in Lecture Notes in Biomathematics, Springer–Verlag, Heidelberg 54: 368–389.

Kareiva, P. 1985. Patchiness, dispersal, and species interactions: consequences for communities of herbivorous insects, in Community Ecology, J. Diamond and T. Case, editors, Harper and Row, New York, pp. 192–206.

Kareiva, P. 1987. Habitat fragmentation and the stability of predator–prey interactions. Nature 321: 388–391.

Kareiva, P. and G. Odell. 1987. Swarms of predators exhibit ʹpreytaxisʹ if individual predators use area restricted search. Am. Naturalist, 130: 233–270.

Keller, E.F. and L.A. Segel. 1970. Traveling bands of chemotactic bacteria: a theoretical analysis. J. Theoretical Biol. 30: 235–248.

Levin, S.A. 1974. Dispersion and population interactions. Am. Naturalist 108: 207–228.

Levin, S. 1981. The role of theoretical ecology in the description and understanding of populations in heterogeneous environments. Amer. Zool. 21: 865–875.

Levin, S. and L. Segel. 1985. Pattern generation in space and aspect. SIAM Review 27: 45–67.

May, R.M. 1978. Host–parasitoid systems in patchy environments: a phenomena–logical model. J. Anim. Ecol. 47: 833–844.

Maynard–Smith, J. 1974. Models in Ecology. Cambridge University Press, Cambridge.

McEvoy, P.B. and C.S. Cox. 1986. Wind dispersal distances in dimorphic achemes of ragwort Senecio jacobaea. Ecology, 68: 2006–2015.

Meinhardt, H. 1982. Models of biological pattern formation. Academic Press, N.Y., 230 pages.

Mimura, M. 1984. Spatial distribution of competing species. In Mathematical Ecology, edited by S. Levin. Lect. Notes in BioMathematics vol. 54: 492–501.

Mollison, D. 1977. Spatial contact models for ecological and epidemic spread. Journal of the Royal Statistical Society, series. B 39: 283–326.

Moore, P.G. 1954. Spacing in plant populations. Ecology 35: 222–227.

Nakamuta, K. 1986. Behavioral mechanisms of switchover in search behavior of the ladybeetle, Coccinella septempunctata. J. Ins. Physiology, in press.

Okubo, A. 1980. Diffusion and ecological problems: mathematical models. Springer–Verlag, Berlin.

Pacala, S.W. and J.A. Silander, Jr. 1985. Neighborhood models of plant population dynamics. I. Single–species models of annuals. Am. Naturalist 125: 385–411.

Price, P., Slobodchikoff, C. and W. Gaud. 1984. A New Ecology: Novel Approaches to Interactive Systems. John Wiley & Sons, New York.

Reynolds, G.D. and L.C. Bliss. 1988. Microenvironmental investigations of tephra covered surfaces at Mount St. Helens. In S.A.C. Keller (ed.), Mount St. Helens: Five years later. Symposium proceedings, Eastern Washington University, Cheney, Washington. In press.

Schaffer, W.M. and E.G. Leigh. 1976. The prospective role of mathematical theory in plant ecology. Systematic Botany 1: 209–232.

Schrodter, H. 1960. Dispersal by air and water – the flight and landing. Pp. 169–227 in Horsfall, J.G. and A.E. Dimond, eds. Plant pathology, an advanced treatise. Vol. 3, The diseased population, epidemics and control. Academic Press, London.

Segel, L. and J. Jackson. 1972. Dissipative structure: an explanation and an ecological example. J. Theor. Biology 37: 545–559.

Sheldon, J.C. and F.M. Burrows. 1973. The dispersal effectiveness of the achene–pappus unit of selected Compositae in steady winds with convection. New Phytologist 72: 665–675.

Skellam, J.G. 1951. Random dispersal in theoretical populations. Biometrika 38: 196–218.

Slatkin, M. and D.J. Anderson. 1984. A model of competition for space. Ecology 65 (6): 1840–1845.

Sutton, O.G. 1953. Micrometerology. R. Krieger, Malaber, Florida.

Thompson, H.R. 1956. Distribution of distance to nth nearest neighbor in a population of randomly distributed individuals. Ecology 37: 391–394.

Venable, D.L. and D.A. Levin. 1983. Morphological dispersal structure in relation to growth habit in the Compositae. Plant Systematics and Evolution 143: 1–16.

Verkaar, H.J., Schenkeveld, A.J. and M.P. van de Klashorst. 1983. The ecology of short–lived forbs in chalk grasslands: dispersal of seeds. New Phytologist 95: 335–344.

Weinberger, H.F. 1978. Asymptotic behavior of a model in population genetics. Pp. 47–96 in J.M. Chadam, ed. Nonlinear partial differential equations and applications. Lecture Notes in Mathematics, Vol. 648. Springer–Verlag, Berlin.

CHAPTER 5

Interactions Between Environment and Competition:

How Fluctuations Mediate Coexistence and

Competitive Exclusion.

Peter L. Chesson

Department of Zoology
Ohio State University
1735 Neil Avenue
Columbus, Ohio 43210

1.    INTRODUCTION

One of the major theoretical and empirical challenges in ecology today is elucidating the role of various kinds of heterogeneity, such as environmental fluctuations, in the dynamics of populations and the organization of communities. There is substantial evidence that stochastic environmental fluctuations have a strong role in population and community processes (Andrewartha and Birch 1954, 1984, Hutchinson 1961, Sale 1977, Connell and Sousa 1983, Sale and Douglas 1984, Grubb 1977, Wiens 1977, 1986, Murdoch 1979, Underwood and Denley 1984, Simberloff 1984, Murdoch et al 1985, Strong 1986, Victor 1986), and as a consequence, a variety of verbal theories of community structure in a stochastic environment have been developed (Hutchinson 1951, 1961, Paine and Vadas 1969, Sale 1977, Grubb 1977, Wiens 1977, Connell 1978). However, the mathematical theories of stochastic environments have not provided an adequate alternative to the classical theory based on deterministic models. There seem to be two reasons for this. First, the existing stochastic models have generally not provided the sort of simple quantitative results that often follow from deterministic models. Second, there is a widespread perception that models incorporating temporal variability may give unfathomable or inconsistent results (Hastings and Caswell 1979, Levin et al 1984). Indeed, an examination of models of communities of competitors reveals that a stochastic environment can have essentially any effect depending on the specific model involved and the assumptions that are made about it. Some models say that environmental variability promotes coexistence (Chesson and Warner 1981, Abrams 1984, Ellner 1984, Shmida and Ellner 1985), others say that environmental variability has little effect on coexistence (Turelli 1981, Chesson and Warner 1981), while others say that environmental variability promotes competitive exclusion (Chesson and Warner 1981).

Using a general model, Chesson (1986) shows that much of the confusion surrounding stochastic models is related to different implicit assumptions about the life–history traits of the species. It is our purpose here to consider a more interpretable formulation of the model of Chesson (1986), and to extend its results. With this new formulation of the general model, it is possible to show how the effects of environmental fluctuations follow from qualitative features of biology. In particular, the life–history traits of model organisms will often dictate an interaction between environmental and competitive effects that determines how environmental variability influences competitive interactions and their outcomes. Moreover, within this framework simple interpretable results are available.

2.    The Model

Consider a community of  n  competing species. To describe their dynamics in discrete time we use the equation

$$X_i(t+1) = G_i(\mathcal{E}_i, C_i)X_i(t), \tag{1}$$

where $X_i(t)$ is the population size of species i at time t, $\mathcal{E}_i$ is an environmentally–dependent parameter of the population of species i, $C_i$ measures the amount of competition that the population is exposed to, and the function $G_i$ converts the environmental and competitive conditions, as measured by the quantities $\mathcal{E}_i$ and $C_i$, into the finite rate of increase for the time interval (t,t+1). The environmentally–dependent parameter will often be something like a density–independent birth rate, survival rate, or seed germination rate. We will assume that the environment fluctuates over time, which means that $\mathcal{E}_i$ fluctuates over time also and can be written as $\mathcal{E}_i(t)$ to emphasize this.

The competitive factor will measure shortage of a critical resource. Sometimes, for instance, it will be density of individuals competing for this resource. In general, $C_i$ will be a function of the species densities in the system and also of their environmentally–dependent parameters, i.e, it will be written in the form

$$C_i = c_i(\mathcal{E}_1,X_1,\mathcal{E}_2,X_2,...,\mathcal{E}_n,X_n), \tag{2}$$

where $c_i$ is some function. Since the $\mathcal{E}_i$'s and the $X_i$'s will be functions of time, so will $C_i$. Moreover, $C_i$ and $\mathcal{E}_i$ will be correlated. If a species is at zero density, it will be assumed that its environmentally–dependent parameter does not contribute to any of the $C_i$'s.

Generally $\mathcal{E}_i$ will be defined so that a larger value reflects more favorable environmental conditions for a species. For example, when the environment causes fluctuating mortality rates, $\mathcal{E}_i$ will be defined as the survival rate. The competitive factor will often increase as the environmentally–dependent parameters increase. For example, if an environmental change increases seed germination then there will be more seedlings around later to compete for moisture, light and nutrients. In some cases an increase in the environmentally–dependent parameter may be a consequence of a greater availability of limiting resources, in which case this assumption is inapplicable and the reverse assumption is more appropriate. However, in all that follows we shall assume that the competitive parameters increase as the environmentally–dependent parameters increase.

Models of population dynamics in a temporally variable environment are best viewed on a log scale for which changes in population size can be written

$$\ln X_i(t+1) - \ln X_i(t) = g_i(\mathcal{E}_i, C_i),\tag{3}$$

where $g_i = \ln G_i$. Note that changes in log population size are found by summing $g_i$ over time, i.e., they are additive in $g_i$. The function $g_i$ can be thought of as an instantaneous per–capita growth rate applicable for the period $(t,t+1)$. We shall refer to it simply as the growth rate.

While change in log population size is additive over the growth rate, $g_i$, the growth rate itself can be additive or nonadditive in its arguments. In the additive case

$$g_i(\mathcal{E}_i, C_i) = A_i(\mathcal{E}_i) + B_i(C_i),\tag{4}$$

for some functions $A_i$ and $B_i$. This additive form applies whenever the quantity

$$\gamma \overset{def}{=} \frac{\partial^2 g_i}{\partial \mathcal{E}_i \partial C_i}\tag{5}$$

is equal to zero. When this quantity is negative, the growth rate is called subadditive, and when positive it is superadditive. These different possibilities for the growth rate are illustrated in figure 1.[*] In the subadditive case, the slope of $g_i$ as a function of $C_i$ is less in a poorer environment, i.e., when the environment is bad the additional effect of competition is slight. Essentially, the growth rate is buffered against simultaneous poor conditions such as a poor environment and high competition. As we shall see later, simple features of biology can produce these buffered growth rates and they are likely to be widespread in nature. The superadditive case is the opposite of a buffer. There the additional effect of competition is worse in the poor environment. The poor environment amplifies the effect of competition.

To analyze the model, we need to introduce some assumptions. We first consider a case where the species have similar biology (we shall call it the symmetric case). The advantage of this case is that results can be obtained for general $g_i$, and general $\mathcal{E}_i$, and it appears to indicate the sort of results to expect in asymmetric settings. We then go on to approximations that allow direct conclusions about asymmetric situations.

## II.1  SYMMETRIC CASE

The first of the symmetry assumptions is that the species have the same competitive factors $(C_1 = C_2 = ... = C_n = C$, say), and the same functions $G_i$ and $g_i$ converting environmental and competitive effects into growth rates. The competitive factor $C$ will

---

* SEE PAGE 71

depend symmetrically on the different species in the system, and we assume that the environmental parameters undergo fluctuations of a similar sort: we assume that the probability distribution of $(\mathcal{E}_1, \mathcal{E}_2, ..., \mathcal{E}_n)$ is exchangeable, i.e., invariant under arbitrary permutation of subscripts. We will, however, assume that the environmentally–dependent parameters do not exhibit perfectly synchronous fluctuations: $P(\mathcal{E}_i = \mathcal{E}_j) < 1$. Justification for such assumptions comes from the fact that some ecologically similar species nevertheless exhibit marked relative fluctuations in their environmentally–dependent parameters. For example, this is the case with tropical trees (Leigh 1982, Connell, unpublished data), and desert annuals (Shmida and Ellner 1984). The final assumption we make is that the vector of environmentally–dependent parameters satisfies the standard random environment assumption: the sequence $[\mathcal{E}_1(0), \mathcal{E}_2(0), ..., \mathcal{E}_n(0)]$, $[\mathcal{E}_1(1), \mathcal{E}_2(1), ..., \mathcal{E}_n(1)]$, ... is an independent and identically distributed sequence of random vectors.

Before analyzing the stochastic case of this model, it is instructive to examine its behavior in a constant environment. The symmetry assumptions imply that in a constant environment there is a neutral equilibrium. To see this, we define the equilibrium amount of competition, $C^*$, by the equation

$$g(\mathcal{E}, C^*) = 0, \tag{6}$$

where $\mathcal{E}$ is the common constant value of the environmentally–dependent parameter. In terms of $C^*$, the set of neutrally stable equilibrium densities is the set $N$ of all points $(X_1, ..., X_n)$ such that

$$c(\mathcal{E}, X_1, ..., \mathcal{E}, X_n) = C^*. \tag{7}$$

This set $N$ will often be a global attractor of the system, but in cases where $g(\mathcal{E}, C)$ can be too large in absolute magnitude, the system will not approach $N$ but will oscillate about it as a consequence of the timelags inherent in this discrete–time formulation.

Except for unlikely functions c, the set N will include points on all faces and edges of the positive cone, $[0, \infty) \times [0, \infty) \times ... \times [0, \infty)$, excluding the origin. Thus N contains points for which $1, 2, ..., n-1$ of the species are extinct in all possible species combinations. Moreover, N will in most cases be a connected set and the equilibria will form a continuum between these states of extinction. Classically, when faced with a deterministic model of this sort ecologists have concluded that only one species can persist when the likely effects of a stochastic environment are taken into account. The reason for this conclusion is the argument that environmental perturbations will cause a random walk to take place in which eventually all but one species becomes extinct. Our analysis below supports this conclusion in only a narrow range of circumstances.

To analyze the model in a stochastic environment we use the invasibility criterion Turelli (1978). The idea is to examine the mean low–density growth rate of each species in the system to see whether species tend to increase or decrease at low density. The mean low–density growth rate of species i is defined as

$$\Delta_i = Eg_i(\mathcal{E}_i, C_i),\tag{8}$$

where the expectation is evaluated under the assumption that $X_i(t) = 0$, and that the remaining species in the system have converged to a unique stationary distribution. The method breaks down if the other species do not converge to a unique stationary distribution. Such convergence is to be expected in regular cases (Revuz 1975), but there are few practical results that can identify when this is true in population models (Ellner 1984). Positive values of the quantity $\Delta_i$ mean that species i tends to increase from low density. Indeed, the strong law of large numbers can generally be invoked to show that $X_i(t)$ will increase, in a fluctuating manner, at least until the approximation (8) to the mean growth rate breaks down (see Chesson 1982, Ellner 1984, Chesson and Ellner 1988). In broad two–species settings it has been shown that positive $\Delta_i$'s imply coexistence in the sense that both species are stochastically boundedly persistent (Chesson and Ellner 1988).

Cases where the $\Delta_i$ are negative mean that a species at low density will have a tendency to converge to extinction. In the general two–species model of Chesson and Ellner (1988) and in the lottery model (Chesson 1982), clear conclusions are available: If $\Delta_1$ and $\Delta_2 < 0$, there is probability 1 that exactly one of the species converges to extinction. If $\Delta_i > 0$, but $\Delta_j < 0$, species j will converge to extinction, and species i will persist in the system.

To apply the invasibility criterion to the general model in this article we focus on a pair of species i and j and define

$$h(\mathcal{E}_i, \mathcal{E}_j) = E[g(\mathcal{E}_i, C) \mid \mathcal{E}_1, ..., \mathcal{E}_n],\tag{9}$$

where species i is at zero density and the other species are at their joint stationary distribution. Expression (9) depends on the environmentally–dependent parameters of all species in the community, but for simplicity of notation only the dependence on the critical pair $\mathcal{E}_i$ and $\mathcal{E}_j$ is represented on the LHS. Note that expression (9) is an average of the growth rate over the species densities in the community for fixed values of the environmentally–dependent parameters. This average is performed using the conditional distribution of $(X_1(t), ..., X_n(t))$ given $(\mathcal{E}_1(t), ..., \mathcal{E}_n(t))$. However, independence of the environment over time implies that $(X_1(t), ..., X_n(t))$ and $(\mathcal{E}_1(t), ..., \mathcal{E}_n(t))$ are independent, and therefore the conditional distribution is the same as the unconditional (marginal) distribution.

The function h itself can be classified as additive, subadditive or superadditive on the basis of the sign of

$$\gamma * (\mathcal{E}_i, \mathcal{E}_j) \overset{\text{def}}{=} \frac{\partial^2 h}{\partial \mathcal{E}_i \, \partial \mathcal{E}_j} . \tag{10}$$

(Note that $\gamma *$ depends on $\mathcal{E}_1, ..., \mathcal{E}_n$, even though only the two focal parameters are represented explicitly as arguments.) Moreover, additivity, subadditivity and superadditivity are properties of $h$ that are inherited directly from $g$. Under the assumption $C$ is increasing in all of its arguments, the sign of $\partial^2 g / \partial \mathcal{E}_i \partial \mathcal{E}_j$ is the same as the sign of $\gamma \, (= \partial^2 g / \partial \mathcal{E}_i \partial C)$. Since $\gamma *$ is simply an average of $\partial^2 g / \partial \mathcal{E}_i \partial \mathcal{E}_j$ over the species densities, $\gamma *$ inherits the sign of $\partial^2 g / \partial \mathcal{E}_i \partial \mathcal{E}_j$ provided the sign of the latter is constant. Otherwise $\gamma *$ inherits the average sign of $\partial^2 g / \partial \mathcal{E} \partial \mathcal{E}_j$. Thus, when we speak of additivity, subadditivity, or superadditivity, it does not matter greatly whether it is defined in terms of $g$ or $h$.

In terms of $h$, the mean low–density growth rate of species $i$ can be evaluated by the formula

$$\Delta_i = Eh(\mathcal{E}_i, \mathcal{E}_j). \tag{11}$$

In addition, note that $h(\mathcal{E}_j, \mathcal{E}_j) = E[g(\mathcal{E}_j, C_j) \mid \mathcal{E}_1, ..., \mathcal{E}_n]$, for the community at its stationary distribution in the absence of species $i$. In this situation, $E \ln X_j(t+1) = E \ln X_j(t)$, and so

$$Eh(\mathcal{E}_j, \mathcal{E}_j) = 0. \tag{12}$$

These results can now be used to decide coexistence using the relationship

$$h(\mathcal{E}_i, \mathcal{E}_i) + h(\mathcal{E}_j, \mathcal{E}_j) - h(\mathcal{E}_i, \mathcal{E}_j) - h(\mathcal{E}_j, \mathcal{E}_i)$$

$$= \int_{\mathcal{E}_j}^{\mathcal{E}_i} \int_{\mathcal{E}_j}^{\mathcal{E}_i} \gamma^* (e_1, e_2) \, de_1 de_2 . \tag{13}$$

Taking expected values in (13), and using (11), (12) and the symmetry assumptions, we get

$$\Delta_i = -\tfrac{1}{2} E \int_{\mathcal{E}_j}^{\mathcal{E}_i} \int_{\mathcal{E}_j}^{\mathcal{E}_i} \gamma * (e_1, e_2) \, de_1 de_2 . \tag{14}$$

If $\gamma *$ is constant in sign it follows that $\Delta_i$ has the opposite sign. Thus, in the subadditive case, the $\Delta$'s of all species in the system will be positive, and the species will coexist by the invasibility criterion. Clearly the value of $\Delta_i$ will tend to be greater for distributions in which $\mathcal{E}_i$ and $\mathcal{E}_j$ have greater probability of being far apart. Thus large values of $\Delta_i$ should be associated with distributions having large variances and negative correlations between $\mathcal{E}_i$ and

$\mathcal{E}_j$. However, $\Delta_i$ will still be positive in all situations where the correlation between $\mathcal{E}_i$ and $\mathcal{E}_j$ is any value less than $+1$. Thus coexistence by this mechanism can still occur even when the environmentally–dependent parameters of the species have strong positive correlations.

In the superadditive case, the opposite sorts of result occur: the $\Delta$'s will be negative and a species at low density will have an average tendency to decrease to extinction. Although there is as yet no general proof, this sort of situation tends to imply that all but one species will ultimately converge to extinction, and the identiy of the surviving species will be a matter of chance. Chesson and Ellner (1988) show that this is so for a class of two–species models.

In the remaining case (additive growth rates) all the mean low–denisty growth rates are zero. The invasibility criterion implies that they have neither an average tendency to increase or decrease at low density. In this case, the growth rates must take the form

$$g_i(\mathcal{E}_i, C_i) = A(\mathcal{E}_i) + B(C), \tag{15}$$

which means that the change in $(\ln X_i - \ln X_j)$ from time $t$ to time $t + 1$ is equal to $A(\mathcal{E}_i) - A(\mathcal{E}_j)$, i.e. it is simply a random variable with mean 0. Thus $\ln X_i - \ln X_j$ undergoes a mean–zero random walk, with no density or frequency dependent effects. Thus, over time, the relative abundances of different species approach extreme values. And in particular, recovery from a state of very low relative abundance is an event with infinite expected waiting time, according to the theory of random walks (Feller 1971). It is reasonable to assume that competition prevents any species from spending correspondingly large periods of time at absolute abundances that are extremely high, and this means that the predicted long periods of time at arbitrarily low relative abundances imply long periods of time at low absolute abundances. As explained for the lottery model (Chesson and Warner 1981), this must be interpreted in the real–world as meaning that the system simplifies to a single species.

In this additive case it is easy to see the consequence of removing some of the symmetry assumptions. We need not assume any symmetry for the distribution of the environmentally–dependent parameters, and we can modify the growth rate (15) so that it takes the form

$$g_i(\mathcal{E}_i, C_i) = A_i(\mathcal{E}_i) + b_i B(C). \tag{16}$$

In this form, competition affects different species to different degrees, but the effects of competition are proportional, reflecting a single underlying competitive factor $C$. It follows that changes in $(\ln X_i)/b_i - (\ln X_j)/b_j$ from $t$ to $t + 1$ are given by the random variable $A_i(\mathcal{E}_i)/b_i - A_j(\mathcal{E}_j)/b_j$. If this random variable always has mean 0, then a mean–zero random walk occurs once more, and the system must be interpreted as simplifying to a single–species. On the other hand, if one species has a larger value of $EA_i(\mathcal{E}_i)/b_i$ than other species, the strong law of large numbers implies that $(\ln X_i)/b_i - (\ln X_j)/b_j$ must converge to $\infty$ for all $j \neq$ i. This means of course that species i drives all the other species asymptotically to extinction, with probability 1.

In summary, the analyses in this section show that environmental variability should promote coexistence when growth rates are subadditive, should promote competitive exclusion when growth rates are superadditive, and in the additive case environmental variability should not alter the conclusion of equilibrium models that different species cannot stably coexist when limited by a common competitive factor.

## II.2   ASYMMETRIC CASES

In nonadditive cases, deviations from the strict symmetry assumptions present complex problems for analysis. It is clear, however, that the conclusions will not collapse with small deviations from strict symmetry because given enough smoothness in the growth rates g, stationary distributions of residents, and mean low–density growth rates, $\Delta_i$, will vary continuously with the joint distribution of the environmentally–dependent parameters. It follows that small asymmetries in the distributions of the environmentally–dependent parameters (favoring some species over others) will not alter the sign of the mean growth rates, except in the additive case for which the effects of asymmetry are well understood, as discussed above.

In general, however, it can be expected that sufficiently large departures from symmetry, in directions that give average advantages to some species over others, will destroy the conclusions of the symmetric case. In particular, in the subadditive case, it can be expected that there will be a relationship between the amount of environmental variability needed for coexistence and the extent to which the environment confers average advantages on some species.

For the subadditive case, expression (14) suggests that the mean low–density growth rates should be increasing functions of the variation in the environmentally–dependent parameters. Hence we would expect that the degree of asymmetry possible before the signs of the $\Delta_i$ disagree with the predictions of the symmetric case, should increase with the amount of variation in the environmentally–dependent parameters. Can we verify these speculations and obtain a quantitive assessment of the joint action of variability and asymmetry? We can do this in the case where the amount of variation is small.

Let $\mathcal{E}^*$ be any particular value of the environmentally–dependent parameter. We consider the effects of environmental variation in the vicinity of this arbitrary fixed value. Since $g(\mathcal{E},C)$ is a decreasing function of $C$, there will be a unique value $C^*$ of $C$ such that $g(\mathcal{E}^*,C^*) = 0$. To obtain general results on the effects of environmental variability we introduce a standard parameterization of the model, which involves transforming the scales of measurement of the environmentally–dependent and competitive parameters. The standard parameters replacing the original parameters $\mathcal{E}$ and $C$ are

$$\mathcal{E}_s = g(\mathcal{E},C^*) \quad \text{and} \quad C_s = -g(\mathcal{E}^*,C). \tag{17}$$

The formula for the growth rate in terms of the new parameters is defined implicitly by the equation

$$g_s(\mathcal{E}_s,C_s) = g(\mathcal{E},C). \tag{18}$$

Note that this parameterization of the model is characterized by the property that

$$g_s(\mathcal{E}_s,C_s) = \mathcal{E}_s - C_s + \Gamma(\mathcal{E}_s,C_s) \tag{19}$$

where the function $\Gamma$ satisfies the equations $\Gamma(0,0) = \partial\Gamma(\mathcal{E}_s,0)/\partial\mathcal{E}_s = \partial\Gamma(0,C_s)/\partial C_s = 0$.

The standard parameterization thus enables the model to be expressed simply in terms of its additive and interactive components. It is important to note that the type of interaction (additive, subadditive or superadditive) is unaffected by this reparameterization because the sign of the cross–partial derivative, $\gamma$, is not altered. From now on we will asssume that the transformation of the model to standard form has been done, and so we will drop the subscript s. Note that in standard form $\mathcal{E}^*$ and $C^*$ are both 0.

Equation (19) implies that a model in standard form can be approximated about the value $\mathcal{E} = C = 0$ by the second–order Taylor expansion

$$g(\mathcal{E},C) = \mathcal{E} - C + \gamma_0 \mathcal{E}C, \tag{20}$$

where $\gamma_0 = \gamma(0,0) = \partial^2 g/\partial\mathcal{E}\partial C$ , at $\mathcal{E} = C = 0$.

Now consider the usual resident–invader situation with two species. The species will be assumed to have the same growth rate function g and the same competitive factor, but the distributions of the environmentally–dependent parameters will be assumed different. In particular, it will be assumed that they have different means. However, in order to use the approximation (20) we must assume that these distributions are both concentrated near the value 0.

Applying (20) to a resident at its stationary distribution gives

$$E\mathcal{E}_j - EC^j + \gamma_0 E\mathcal{E}_j C^j = 0 \tag{21}$$

where $C^j$ is random variable representing the competition variable C for the resident at its stationary distribution. It follows from this that

$$E\mathcal{E}_j - EC^j + \gamma_0(E\mathcal{E}_j)(EC^j) = -\gamma_0 \chi, \tag{22}$$

where $\chi$ is the covariance between $\mathcal{E}_j$ and $C^j$. This can be rearranged to give the following formula for $EC^j$:

$$EC^j = \frac{E\mathcal{E}_j + \gamma_0\chi}{1 - \gamma_0 E\mathcal{E}_j}. \tag{23}$$

The mean growth rate of an invader is

$$\Delta_i = E\mathcal{E}_i - EC^j + \gamma_0 E\mathcal{E}_i C^j. \tag{24}$$

To go beyond this formula we need to relate $\mathcal{E}_i$ and $\mathcal{E}_j$, and it turns out that all we need is the regression of $\mathcal{E}_i$ on $\mathcal{E}_j$, i.e. the function $E[\mathcal{E}_i|\ \mathcal{E}_j]$. Provided this regression function is differentiable, we can use the first order Taylor approximation

$$E[\mathcal{E}_i|\ \mathcal{E}_j] = E\mathcal{E}_i + b(\mathcal{E}_j - E\mathcal{E}_j), \tag{25}$$

where b is a constant. This first order approximation is all that is needed to retain the second order character of our approximation for $\Delta_i$. This implies that $E\mathcal{E}_i C^j = E\{E[\mathcal{E}_i|\ \mathcal{E}_j]C^j\} = (E\mathcal{E}_i)(EC^j) + b\chi$. Substituting this and (23) in (24) we obtain

$$\Delta_i = E\mathcal{E}_i - (E\mathcal{E}_j + \gamma_0\chi)\ \frac{1 - \gamma_0 E\mathcal{E}_i}{1 - \gamma_0 E\mathcal{E}_j} + \gamma_0 b\chi \tag{26}$$

Eliminating terms of order higher than 2 (the limit of accuracy of these approximations) this reduces to

$$\Delta_i = (\Delta E\mathcal{E})\ (1+\gamma_0 E\mathcal{E}_j) - \gamma_0(1-b)\chi, \tag{27}$$

where $\Delta E\mathcal{E} = E\mathcal{E}_i - E\mathcal{E}_j$.

This mean growth–rate formula is most easily interpreted when converted into a condition for $\Delta_i$ to be positive, allowing invasion to occur. The condition is

$$E\mathcal{E}_i > E\mathcal{E}_j + \frac{\gamma_0\ (1-b)\chi}{1+\gamma_0 E\mathcal{E}_j}. \tag{28}$$

Note that the regression coefficient b can be written as the product of the correlation coefficient between $\mathcal{E}_i$ and $\mathcal{E}_j$ and the ratio of their standard deviations. In most cases the value of b will be less than 1. It follows that in subadditive situations (negative $\gamma_0$) the invader can increase from low density even though it is at a disadvantage to the resident (has a smaller mean environmental parameter). The disadvantage that species i has to species j can be measured by $E\mathcal{E}_j - E\mathcal{E}_i$, and the supremum of values of this quantity permitting invasion is, by (28), $|\gamma_0|(1-b)\chi/(1+\gamma_0 E\mathcal{E}_j)$. Thus, species i can invade at a greater disadvantage for larger values of $|\gamma_0|$ (the degree of subadditivity), larger values of $\chi$ (the covariance between the environment and competition), and smaller values of the coefficient of regression of the environmental parameter of the invader on that of the resident.

The invader can always successfully invade in this subadditive case if it has an advantage over the resident ($E\mathcal{E}_i > E\mathcal{E}_j$), and so these conditions for invasion of a weaker species are also conditions for coexistence. A multispecies extension of these results is possible but is deferred to a subsequent article.

If $\gamma_0$ is 0 (the additive case), invasion can only occur if the invader is superior on average to the resident, as the earlier exact results demonstrate. With positive $\gamma_0$ (superadditivity), negative mean low–density growth rates are favored by environmental variability for both species of a two–species system. This means that whichever species first fluctuates to low–density is likely to become extinct. However, with enough asymmetry in the system, it is possible for one species to have a positive mean low–density growth rate, while that for other species is negative. Thus, there is a specific winner and loser in competition. To have a positive mean low–density growth rate, species i must be superior to species j by at least a little more than $\gamma_0(1{-}b)\chi/(1{+}\gamma_0 E\mathcal{E}_j)$. Thus conditions unfavorable to invasion by species i are defined by a large interaction, $\gamma_0$, a large covariance, $\chi$, or a small regression coefficient, b.

The covariance, $\chi$, in these results, will vary with the details of the specific situation. It is possible, however, to get some feeling for it by considering the case of strong competition for a resource such as space. If this strong competition tends to lead to all space being filled, then in a single–species system the growth rate, $g(\mathcal{E},C)$, will approach 0. Thus in two–species space–filling models a resident will satify the equation $0 = g(\mathcal{E}_j, C^j)$. It follows that $(\mathcal{E}_j)^2 = (C^j)^2 +$ terms of order higher than 2. Hence

$$V\mathcal{E}_j = VC^j = \chi,$$ to second order, where V means variance. Thus, in such space filling situations, $\chi$ can be approximated by the variance of the environmentally–dependent parameter. In other situations this will not be true. In particular the variance of $C^j$ will have components of variation due to fluctuations in species densities as well as variation in the environmentally–dependent parameter. Thus $VC^j$ may often be larger than $V\mathcal{E}_j$. Nevertheless, the covariance, $\chi$, is still likely to be closely related to $V\mathcal{E}_j$.

The results that we have obtained here extend those of the symmetric model to asymmetric distributions of the environmental parameters. Moreover, here we have obtained explicit quantitative results defining conditions for coexistence and competitive exclusion showing how they are affected by different factors in the system. Assumptions that are retained in this analysis are equality of the competition variables $C_i$ and equality of the function $g_i$ converting the environmental and competitive effects into the growth rate. However, these remaining assumptions also can be removed (Chesson, in prep.)

## III.  HOW ADDITIVITY AND NONADDITIVITY ARISE

Mathematical models using difference equations have focused on organisms with strictly nonoverlapping generations. The best examples of such organisms are univoltine insects and annual plants without a seed bank. In such organisms there will usually be a period of reproduction followed by a period of juvenile growth, mortality, and final maturation to adulthood before the next breeding season. The per–capita number of adults at the next breeding season can be represented as the product of the per–capita birth rate B, for adults alive at the beginning of the breeding season, and the juvenile survival rate $\Theta$, giving the fraction of births that ultimately yield an adult at the beginning of the next breeding season. Both stages may be affected by environment and competition, but consider for the moment an environmental factor that affects only the birth rate B. The quantity B is then an environmentally–dependent parameter, i.e., $\mathcal{E}_i$ = B. If juveniles compete, then $\Theta$ is dependent on $C_i$, the competitive factor for the species, and so we have

$$G_i(\mathcal{E}_i, C_i) = \mathcal{E}_i \Theta(C_i). \tag{29}$$

Since this is a product, $g_i(\mathcal{E}_i, C_i)$ is necessarily additive. Note that it is quite reasonable in this case for $C_i$ to be an increasing function of the environmentally_dependent parameters of the species in the community because a higher birth rate will lead to a larger number of juveniles and therefore more competition among those juveniles. If the factor causing the high birth rate persists during the period of juvenile competition and in some way is able to more than make up for the increased number of juveniles, then the reverse assumption that $C_i$ is a decreasing function of $\mathcal{E}_i$ must be made. Nevertheless, both $g_i(\mathcal{E}_i, C_i)$ and $h(\mathcal{E}_i, \mathcal{E}_j)$ remain additive.

Other variations on the arrangement above retain the same additive property. Instead of the birth rate being variable, or in addition to this, early juvenile mortality could be environmentally–dependent, but later juvenile mortality could depend on competition. Or the birth rates might be affected by competition among adults while the juvenile survival rates are environmentally–dependent. In all cases $g_i$ remains additive.

Another important case is where the environmental factor and the competitive factor operate simultaneously. For example, suppose they operate during a mortality phase. The environmentally–dependent parameter could then be an instantaneous contribution to survival due to the environment while the competitive factor would be the instantaneous contribution to mortality due to competition. The latter could be called $C_i'$, because it would not be equivalent to the total competitive factor for the relevant period. The change in population size during the juvenile period could then be represented as

$$\frac{1}{X_i} \frac{dX_i}{dt} = -C_i' + \mathcal{E}_i \ . \tag{30}$$

Note that $\mathcal{E}_i$ is a negative number. Letting h be the length of the relevant juvenile period, it follows that the fraction of the population surviving over the juvenile period is

$$e^{-C_i + h\mathcal{E}_i},$$
(31)

where

$$C_i = \int_0^h C_i'(u)du .$$
(32)

Note that $C_i'(u)$ will be dependent on the environment and the species densities in the system at the time $u$ and even possibly at previous points in time. However, species densities at time $u$ are functions of what they were at time 0. It follows that the representation (2) for $C_i$ is still applicable.

These results show that organisms with very simple life histories will often have additive growth rates if the environmentally–dependent parameter is something like a birth rate or survival rate. More complex features of biology are necessary before we find nonadditive growth rates and interesting conclusions. It turns out that seemingly minor perturbations of the model can have large effects on these conclusions: additivity is not a structurally stable feature of a model.

### Organisms with nonoverlapping generations

Suppose now that we are dealing with organisms that are iteroparous, i.e., the adults may remain reproductive for several seasons. In many such organisms, reproduction, juvenile survival, and maturation are processes that are much more environmentally and competitively sensitive than adult survival (Chesson 1986). Thus we shall represent adult survival as a constant $s$. The net per–capita number of new adults produced in a year can be represented as a function $G_i'(\mathcal{E}_i,C_i)$, where $\ln G_i'(\mathcal{E}_i,C_i)$

$(= g_i'(\mathcal{E}_i,C_i))$ is additive. It follows that

$$g_i(\mathcal{E}_i,C_i) = \ln\{s + G_i'(\mathcal{E}_i,C_i)\}.$$
(33)

Elementary calculus shows that

$$\gamma = s\alpha\beta G_i'/G_i^2,$$
(34)

where $\alpha = \partial g_i'/\partial\mathcal{E}_i$ and $\beta = \partial g_i'/\partial C_i$. Since $\beta$ is negative, while all other factors in (34) are positive, we see that $\gamma$ is negative and therefore $g_i$ is subadditive. It follows that overlapping generations and iteroparity promote coexistence in a variable environment. It must be kept in mind, however, that we assume positive covariance between environmentally–dependent parameters and competition, and that adult survival is insensitive to environmental and competitive factors.

The discussion above assumes that generations overlap in the adult phase of the life cycle. However, the results apply also to annual plants with seed banks by interpreting $s$ as the fraction of seeds that remain viable but do not germinate. This representation is satisfactory provided germination itself is not the environmentally–dependent parameter. When germination is the environmentally–dependent parameter we obtain the equation

$$g_i(\mathcal{E}_i, C_i) = \ln\{s(1 - \mathcal{E}_i) + G'_i(\mathcal{E}_i, C_i)\}. \tag{35}$$

For this model, we obtain

$$\gamma = s[(1 - \mathcal{E}_i)\alpha + 1]\beta G'_i / G_i^2 . \tag{36}$$

Again $\gamma$ is negative, showing that growth rates are subadditive and that coexistence will be promoted in a variable environment, as first demonstrated in a specific seed–bank model by Ellner (1984).

The superadditive situation is also easily seen in these models. To obtain that case, one can make the survival rate in (33) and (35) the environmentally–dependent parameter. These growth rates become respectively

$$g_i(\mathcal{E}_i, C_i) = \ln\{\mathcal{E}_i + G'_i(\mathcal{E}_i, C_i)\} \tag{37}$$

and

$$g_i(\mathcal{E}_i, C_i) = \ln\{\mathcal{E}_i(1 - \theta) + G'_i(\mathcal{E}_i, C_i)\}, \tag{38}$$

where the germination rate is now $\theta$. In both cases, simple calculus shows that $\gamma$ has the same sign as

$$1 - \alpha \mathcal{E}_i . \tag{39}$$

The quantity $\alpha$ measures the dependence of reproduction and juvenile survival on adult or seed survival, at a fixed level of competition. The existence of tradeoffs between adult survival and reproduction in some organisms (Nur 1984) suggests that $\alpha$ may be negative. In any case, large positive values of $\alpha$ seem unlikely. Thus it appears that (39) will often be positive indicating superadditivity, and competitive exclusion will be promoted by environmental fluctuations.

A model with poor alternative resources.

In foraging theory it is recognized that species may change their selection of resources as the abundances of the resources change. However, classical theories of resource consumption, which receive their most detailed expression in MacArthur's (1970) comsumer–resource model, do not allow for this possibility. Consider a situation in which there are good resources that the organisms need to sustain a positive growth rate and poor resources that are able to prevent precipitous decline of the population but are not able provide positive growth. For example, Scott (1980), and Scott and Murdoch (1983) have found that Notonecta populations show markedly different growth responses to different prey species. Moreover, the relative availabilities of these species vary over time. In classical resource competition theories, limiting resources allow population increase when these resources are sufficiently abundant. Population increase causes decline in availability of limiting resources by increased consumption, which eventually halts population growth. This scenario does not apply to poor resources that merely moderate the rate of decline of a species, because population increase cannot occur in response to such resources. Thus, we will regard availability of the poor alternative resources as insensitive to consumer species density.

Let $s\ (< 1)$ be the finite rate of increase when a species is concentrating on poor resources. These resources set a minimum growth rate for a population and therefore the finite rate of increase of the population might be represented as

$$G_i(\mathcal{E}_i, C_i) = \max\{s, G'_i(\mathcal{E}_i, C_i)\}, \tag{40}$$

where $G'_i(\mathcal{E}_i, C_i)$ is the finite rate of increase applicable in the absence of the these poor alternative resources. This equation assumes that the growth rate on the good resources is environmentally sensitive, reflecting possibly environmental dependence in ability to use the good resources. It assumes also a complete switch from good to poor resources at the point where poor resources actually become more profitable. Such a complete switch is unlikely unless the species change habitat or foraging mode to use the poor resources. However, as we shall see it represents the worst case situation for nonadditive growth rates, and so our analysis of this model is conservative.

If $g'_i(\mathcal{E}_i, C_i)$ is additive, it follows that $\gamma$ is zero everywhere except on the set M of $\mathcal{E}_i$ and $C_i$ values for which $G'_i(\mathcal{E}_i, C_i) = s$. On M, $\gamma$ does not exist. It follows that $\gamma$ cannot characterize the joint dependence of $g_i(\mathcal{E}_i, C_i)$ on $\mathcal{E}_i$ and $C_i$ when $(\mathcal{E}_i, C_i)$ can cross from one side to the other of the set M. However, this problem can be got around using the following finite difference definition of subadditivity:

$$g(\mathcal{E}', C') - g(\mathcal{E}, C) \le [g(\mathcal{E}', C) - g(\mathcal{E}, C)]$$
$$+ [g(\mathcal{E}, C') - g(\mathcal{E}, C)] \tag{41}$$

for $\mathcal{E}' > \mathcal{E}$ and $C' > C$. The growth rate $g_i(\mathcal{E}_i, C_i)$ satisfies (41), and the inequality is strict whenever $(\mathcal{E}', C')$ and $(\mathcal{E}, C)$ are on opposite sides of the curve M in $(\mathcal{E}_i, C_i)$–space. Thus, our model of poor alternative resources is subadditive in this finite–difference sense. It is easy to see using the finite–difference methods of Chesson (1986) that coexistence will be promoted in a fluctuating environment provided the $(\mathcal{E}_i, C_i)$ values of the invader fluctuate on either side of M.

Other versions of this model in which there is a smooth change over from one resource to another, will have $\gamma$ existing everywhere. Generally, $\gamma$ will be negative in the region where both resources are being utilized, and therefore the model will show subadditivity in this range. Coexistence will thus be promoted by environmental fluctuations.

### Fluctuating resource uptake rates.

Abrams (1984) put forward a model in which the resource uptake rates of a species are the environmentally–dependent parameters. He showed that under a variety of conditions these fluctuating uptake rates promoted coexistence. In all cases where coexistence occurred, however, he assumed either overlapping generations or nonlinear conversion of resource into consumer biomass. Here we seek the effects of the fluctuating uptake rates alone without these other complications. Thus we consider the following formula for the finite rate of increase of the population:

$$G_i(\mathcal{E}_i, C) = f(\mathcal{E}_i R(C)) - m_i \ . \tag{42}$$

Here $R(C)$ is the amount of resource present in the environment as a function of the level of competition $C$ for the resource. The environmentally–dependent parameter $\mathcal{E}_i$ is a measure of relative availablility of the resource to consumer species i: it alters the effective abundance of the resource to consumer species i. The functional response f gives the actual per–capita resource consumption as a function of resource availability, $\mathcal{E}_i R(C)$. Finally, $m_i$ represents a minimum amount of resource that must be consumed before any individuals are produced in the next time period. This minimum amount of resource would be 0 if all resource consumption resulted directly in the production of new individuals. For example, this occurs in parasitoids in which the resource is a host population that provides food for developing larvae alone. Note that the model assumes nonoverlapping generations.

In the case where $m_i = 0$, routine calculations show that

$$\gamma = R'(C)\phi'(\mathcal{E}_i R(C)) + \mathcal{E}_i R(C)R'(C)\phi''(\mathcal{E}_i R(C)), \tag{43}$$

where $\phi = \ln f$ and the primes indicate derivatives. Since $R'(C)$ is necessarily negative, it follows that the sign of $\gamma$ is opposite to the sign of $\phi' + x\phi''$, where $x = \mathcal{E}_i R$. By solving the differential equation $\phi' + x\phi'' = 0$, one obtains the following formula for a functional response that will yield additive growth rates:

$$f(x) = f(1)x^{f'(1)/f(1)}. \tag{44}$$

It is not difficult to see that in the subadditive case, $f(x)$ will be greater than the RHS of (44) for all x except $x - 1$; and in the supcradditive case, $f(x)$ will be less than the RHS of (44) everywhere except $x = 1$. Functional responses generally approach a constant value as x increases, which means that they generally do not satisfy the power law relationship implied by (44). Indeed, at least for large x, they will be less than (44). Functional responses may satisfy a power law at low x but since they saturate they may often follow a law that can be approximated by the formula

$$f(x) = \frac{a x^b}{c + ax^b} \ . \tag{45}$$

It is a simple calculus exercise to show that this functional response always gives superaddivitive growth rates. Moreover, the fact that functional responses generally saturate will invariably add an element of superadditivity to the model even if the formula (45) is not a good approximation.

When the value of $m_i$ is positive, simple calculus shows that the sign of $\gamma$ is opposite the sign of

$$\phi' + x\phi'' - x(\phi')^2 \frac{m_i}{1 - m_i} . \tag{46}$$

Hence the maintenance requirement itself is a factor that promotes superadditivity. Indeed, it will lead to superadditivity in instances where the functional response by itself would not, for instance, in cases where the functional response is linear with no upper limit on the consumption rate. In general, this model implies that fluctuating resource consumption rates should favor competitive exclusion, unless there are other factors in the model that oppose this tendency.

Abrams' (1984) model differs from this one in that in all cases he assumed a linear functional response. The presence of a maintenance requirement would imply superadditivity from the results above; however, Abrams is able to get subadditivity by adding overlapping generations or accelerating conversion of resource into consumer biomass. Thus he finds that fluctuating resource–uptake rates promote coexistence. Finally, it should be kept in mind that here consumption rates are assumed to vary as a consequence of fluctuations in the availability of fixed abundances of resource to consumer species. This means that environmental factors not tied to resource abundance interfere with the ability of the species to consume the resource. Different biological assumptions about the nature of variability could lead to different results.

## CONCLUSION

An interaction between the environment and competition is necessary before environmental fluctuations have any overall effect on community structure. In many ways this is hardly surprising: if the environment does not interact with competitive processes then their long–term effect should be moot because the fluctuations should just average out to the equivalence of a constant value. The additive case (no interaction) confirms this intuitive expectation. With negative or subadditive interactions, competitive exclusion is opposed, while with positive or superadditive interactions competitive exclusion is promoted. Within this framework the results are clear and unambiguous. Ambiguity often arises in stochastic models because the space of probability distributions is infinite dimensional, yet discussions of the effects of random variables often focus on just two dimensions, the mean and variance alone. The effects of the neglected parameters are responsible for the ambiguous results. In the investigations here, the results are clear because in the symmetric case the symmetry itself prevented the infinite dimensionality from being expressed. Symmetry in essence led to cancellation of the effects of parameters other than a measure of variation. In the asymmetric analysis, the standard parameterization concentrated all the important effects in the means and covariances. Thus ambiguities due to the action of other parameters did not arise.

While these results are simple and interpretable mathematically, it is important that they are also interpretable biologically in terms of buffers, amplifiers and neutral traits. The subadditive case can be interpreted as the presence of biological properties that buffer unfavorable combinations of events. When the environment is poor for a species, the additional effect of competition is slight. Thus the form of the growth rate provides a buffer against simultaneously having a poor environment and strong competition. Biologically, such buffers can result from overlapping generations or the availability of alternative resources.

The superadditive case involves amplification of unfavorable combinations of events: when environmental conditions are poor the additional effect of competition is strong. Amplification can occur when a feature that would buffer when it is itself insensitive to the environment, is made the environmentally–dependent parameter. For example, competition among juveniles or seedlings will have more effect on the growth rate at times of poor adult or seed survival, and so environmental sensitivity of adult and seed survival is a feature that leads to amplification of the combined effects of a poor environment and strong competition. We found also that the shape of the functional response can lead to superadditivity when resource uptake rates fluctuate.

Additive growth rates result from neutral traits. But we found that neutral traits can easily be overcome by other traits that act as buffers or amplifiers. Thus, strictly additive growth rates must be regarded as idealizations that may be approximated in nature but are probably never actually achieved. Neutrality seems to arise generally as a result of sequential action of environmental and competitive effects in organisms that have strictly nonoverlapping generations. Simple modeling approaches often treat events as taking place sequentially, not interacting simultaneously, and these approaches often lead to additive growth rates. While such approaches have been adequate for consideration of equilibrium situations, our results show that the addition of just a little more reality can markedly change the conclusions in nonequilibrium settings.

Acknowledgements

I am grateful for comments on the manuscript by Peter Abrams and Joe Connell. This work was aided by NSF grant BSR–8615028.

REFERENCES

Abrams, P. 1984. Variability in resource consumption rates and the coexistence of competing species. Theoret. Pop.Biol. 25: 106–124.

Andrewartha, H. G., and L. C. Birch. 1954. The Distribution and Abundance of Animals. Chicago University Press, Chicago.

Andrewartha, H. G., and L. C. Birch. 1984. The Ecological Web: More on the
    Distribution and Abundance of Animals. Chicago University Press, Chicago. 566 pp.

Armstrong, R. A., and R. McGehee. 1976. Coexistence of species competing for shared
    resources. Theoret. Pop. Biol. 9:317–328.

Chesson, P. L. 1982. The stabilizing effect of a random environment. J. Math. Biol.
    15: 1–36.

Chesson, P. L. 1986. Environmental variation and the coexistence species. pp. 240–256 in
    T. Case and J. Diamond, eds, Community Ecology, Harper and Row.

Chesson, P. L. 1987. The relationship between life–history traits and the coexistence of
    competitors in a fluctuating environment. In preparation.

Chesson, P. L., Ellner, S. P. 1988. Invasibility and stochastic boundedness in monotonic
    competition models. J. Math. Biol., in press.

Chesson, P. L., and R. R. Warner. 1981. Environmental variability promotes coexistence
    in lottery competitive systems. Amer. Natur. 117: 923–943.

Connell, J. H. 1978. Diversity in tropical rainforests and coral reefs. Science
    199: 1302–1310.

Connell, J. H., and W. P. Sousa. 1983. On the evidence needed to judge ecological
    stability or persistence. Amer. Natur. 121:789–824.

Ellner, S. P. 1984. Stationary distributions for some difference equation population
    models. J. Math. Biol. 19: 169–200.

Ellner, S. and A. Shmida. 1981. Why are adaptations for long–range seed dispersal rare in
    desert plants? Oecologia 51: 133–144.

Feller, W. 1971. An Introduction to Probability Theory and its Application, Vol. I. 2d ed.
    Wiley, N.Y.

Grubb, P. J. 1977. The maintenance of species richness in plant communities: the
    regeneration niche. Biol. Rev. 52: 107–145.

Hastings, A., Caswell, H. 1979. Role of environmental variability in the evolution of
    life history strategies. Proc. Natl. Acad. Sci. USA 76: 4700–4703.

Hubbell, S. P. 1980. Seed predation and the coexistence of tree species in tropical
    forests. Oikos 35: 214–299.

Hutchinson, G. E. 1951 Copepodology for the ornithologist. Ecology 32: 571–577.

Hutchinson, G. E. 1961. The paradox of the plankton. Amer. Natur. 95: 137–145.

Leigh, E. G. Jr. 1982. Introduction: why are there so many kinds of tropical trees?
    pp. 63–66 in E.G. Leigh, Jr., A. S. Rand, and D. W. Windsor, eds, The Ecology of a
    Tropical Forest.

Levin, S. A., Cohen, D., and Hastings, A. 1984. Dispersal strategies in patchy
    environments. Theoret. Pop. Biol. 26, 165–191.

MacArthur, R. H. 1970. Species packing and competitive equilibrium for many species.
    Theoret. Pop. Biol. 1, 1–11.

May, R. M. 1974. On the theory of niche overlap. Theor. Pop. Biol. 5: 297–332.

Murdoch, W. W. 1979. Predation and the dynamics of prey populations. Fortschr. Zool. 25: 245–310.

Murdoch, W. W., J. Chesson and P. L. Chesson. 1985. Biological control in theory and practice. Amer. Natur. 125, 344–366.

Nur, N. 1984. The consequences of brood size for breeding blue tits I. Adult survival, weight change and the cost of reproduction. J. Anim. Ecol. 53, 479–496.

Paine, R. T., and R. L. Vadas. 1969. The effect of grazing by sea urchins Strongylocentrus spp, on benthic algal populations. Limnol. Oceanogr. 14: 710–719.

Revuz, D. 1975. Markov Chains. North Holland, Amsterdam. 336 pp.

Sale, P. F. 1977. Maintenance of high diversity in coral reef fish communities Amer. Natur. 111: 337–359.

Sale, P. F., Douglas, W. A. 1984. Temporal variability in the community structure of fish on coral patch reefs and the relation of community structure to reef structure. Ecology 65, 409–422.

Scott, M. A. 1980. The effect of the general predator Notonecta (Hemiptera: Notonectidae) on community structure: selective predation and the effect of alternative prey on predator dynamics. Dissertation. University of California, Santa Barbara, California, U.S.A.

Scott, M. A., and W. W. Murdoch. 1983. Selective predation by the backswimmer Notonecta. Limnology and Oceanography 28, 352–366.

Shmida, A., and S. P. Ellner. 1985. Coexistence of plant species with similar niches. Vegetatio 58, 29–55.

Simberloff, D. 1984. Properties of coexisting bird species in two archipelagoes. In D. R. Strong, Jr., D. Simberloff, L. G. Abele, A. B. Thistle, eds., Ecological Communities: Conceptual Issues and the Evidence, pp. 234–253.

Strong, D. R. 1986. Density vagueness: abiding the variance in the demography of real populations. pp. 257–268 in J. Diamond, T. J. Case, eds., Community Ecology, Harper and Row.

Turelli, M. 1978. Does environmental variability limit niche overlap? Proc. Natl. Acad. Sci. USA 75: 5085–5089.

Turelli, M. 1981. Niche overlap and invasion of competitors in random environments I. Models without demographic stochasticity. Theoret. Pop. Biol. 20: 1–56.

Underwood, A. J. and E. L. Denley. 1984. Paradigms, explanations, and generalizations in models for the structure of intertidal communities on rocky shores. In D. R. Strong, Jr., D. Simberloff, L. G. Abele, and A. B. Thistle, eds., Ecological Communities: Conceptual Issues and the Evidence, pp. 151–180. Princeton University Press, Princeton.

Victor, B. C. 1986. Larval settlement and juvenile mortality in a recrutiment limited coral reef fish population. Ecol. Mono. 56, 145–160.

Warner, R. R., and P. L. Chesson. 1985. Coexistence mediated by recruitment fluctuations: a field guide to the storage effect. Amer. Natur., 125, 769–787.

Wiens, J. A. 1977. On competition and variable environments. Amer. Scientist 65:590–597.

Wiens, J. A. 1986. Spatial scale and temporal variation in studies if shrubsteppe birds, pp. 154–172 in J. Diamond, T. J. Case, eds., Community Ecology, Harper and Row.

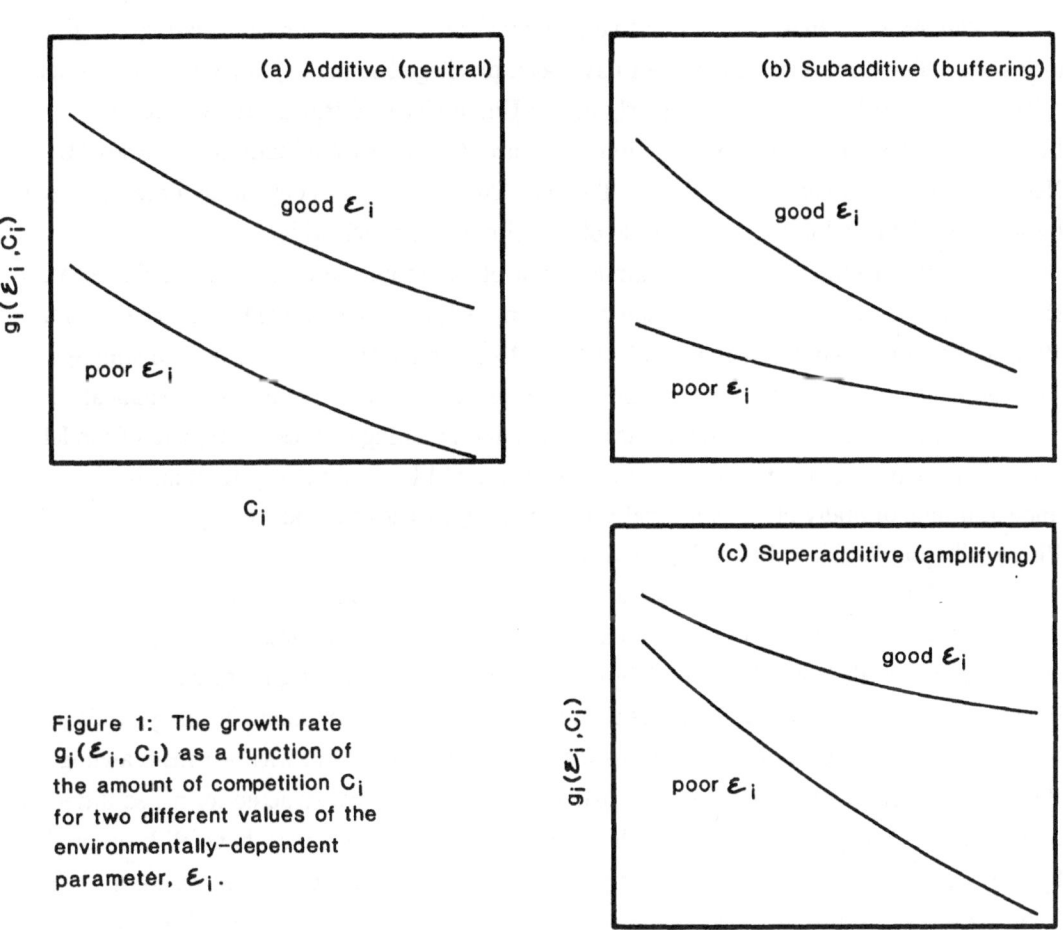

Figure 1: The growth rate $g_i(\varepsilon_i, c_i)$ as a function of the amount of competition $c_i$ for two different values of the environmentally–dependent parameter, $\varepsilon_i$.

# CHAPTER 6

## Untangling 'An Entangled Bank':
## Recent Facts and Theories About Community Food Webs

Joel E. Cohen

Rockefeller University, 1230 York Avenue, New York, NY 10021–6399, U.S.A.

## I.  INTRODUCTION

This paper is an expository and nontechnical review of some recent discoveries about food webs.  The discoveries are those I have been privileged to make jointly with two splendid collaborators:  Frédéric Briand, formerly at the University of Ottawa and now at the International Union for the Conservation of Nature, Gland, Switzerland; and Charles M. Newman, at the University of Arizona, Tucson.  These discoveries depend on data collected by scores of field ecologists, so the circle of contributors is much wider.

I do not attempt here a panoramic review of community ecology (for which, see e.g. Diamond and Case, 1986; Kikkawa and Anderson, 1986), or even of food webs (see Pimm, 1982, and this volume; DeAngelis et al., 1983).  I attempt rather to describe in a simple way some new facts that, in their original presentations, may appear forbiddingly technical.  I hope that, in the future, a thorough theoretical understanding of these facts, and of models that can provide quantitative explanations of them, will lead eventually to quantitative understanding of many other empirically justified approaches to food webs (e.g. Cohen, 1978; Pimm, Chapter 7, this volume; Sugihara, 1984).

Food webs describe which species of organisms in a community eat which other species, if any.  Food webs figure in one of the most famous paragraphs in biology, the last paragraph of Charles Darwin's book, "On the Origin of Species."  That paragraph begins:  "It is interesting to contemplate an entangled bank, clothed with many plants of many kinds, with birds singing on the bushes, with various insects flitting about, and with worms crawling through the damp earth, and to reflect that these elaborately constructed forms, so different from each other, and dependent on each other in so complex a manner, have all been produced by laws acting around us."  Darwin summarizes his theory of evolution and resumes:  "Thus, from the war of nature, from famine and death, the most exalted object which we [Darwin speaks anthropocentrically here] are capable of conceiving, namely, the production of the higher animals, directly follows."  The study of food webs is the study of that war of nature, and of the laws acting around us which govern it.

So far as I know, food webs were first described in scientific detail at the beginning of this century.  Now, more than a century and a quarter after Darwin published his theory of evolution, enough examples of the war of nature have been patiently observed and recorded to make it possible to understand how the lines of battle are drawn.  I will illustrate what a food web is and how a food web is described.

Even relatively simple webs may seem very complex, too complex to understand whole. Until recent decades, ecological theorists studied small components of webs, such as interactions between a predator species and a prey species. I have turned in the opposite direction, in the hope that ensembles or collections of food webs might display simple general properties that are not evident from any single web. This hope, after the long labors of gathering and analyzing data on many webs, has been fulfilled. I will present some quantitative empirical generalizations that we have recently discovered about food webs.

Then I will present two models. One of the models unifies the quantitative generalizations. This successful model is ridiculously simple. Any self−respecting field ecologist would sneer at it. (Rightly so: Where are its dynamics, its spatial structure, its representation of behavior and genetics and physiology and energy flow and environmental fluctuations?) I present it only because no other model at present connects and explains quantitatively what is observed. We call the successful model the cascade model. After showing that the cascade model describes what we know already, I then show that it makes novel predictions about things we did not know already. These predictions can be tested.

Finally, I will outline some potential uses of facts and theories about food webs.

## II.    TERMS

Let me introduce some terms and illustrate them with an example. A food web is a collection of trophic species, together with their feeding relations. A trophic species is a collection of organisms that have the same diets and the same predators. A biological species, in the usual use of the term, refers to a collection of organisms with shared genetics. A trophic species will sometimes be a biological species, but not always. A trophic species may be a biological species of plant or animal, or several species, or a stage in the life cycle of one biological species. Hereafter, the word "species" means "trophic species."

Each arrow in a food web goes from food to eater, or from prey to predator. I call each arrow a "link", short for "trophic link."

Fig. 1 is a picture of the food web on an island in the Pacific Ocean. Some species are top, meaning that no other species in the web eats them, e.g., reef heron, starlings. Notice that the web omits decomposers. Some species are intermediate, meaning that at least one species eats them, and they eat at least one species, e.g., insects, skinks, fish. Some species are basal, meaning that they eat no other species, e.g., algae, phytoplankton. To quantify the structure of webs, we count the numbers of species that are top, intermediate and basal.

These three kinds of species specify four kinds of links: basal−intermediate links, e.g., phytoplankton to zooplankton; basal−top links, e.g., coconut to man; intermediate−intermediate links, e.g., zooplankton to fish; and intermediate−top links, e.g., fish to frigate birds. We also count the numbers of links of each of these four kinds.

Fig.1  Food web in the Kapingamarangi Atoll.  From p. 157 of Niering, 1963.

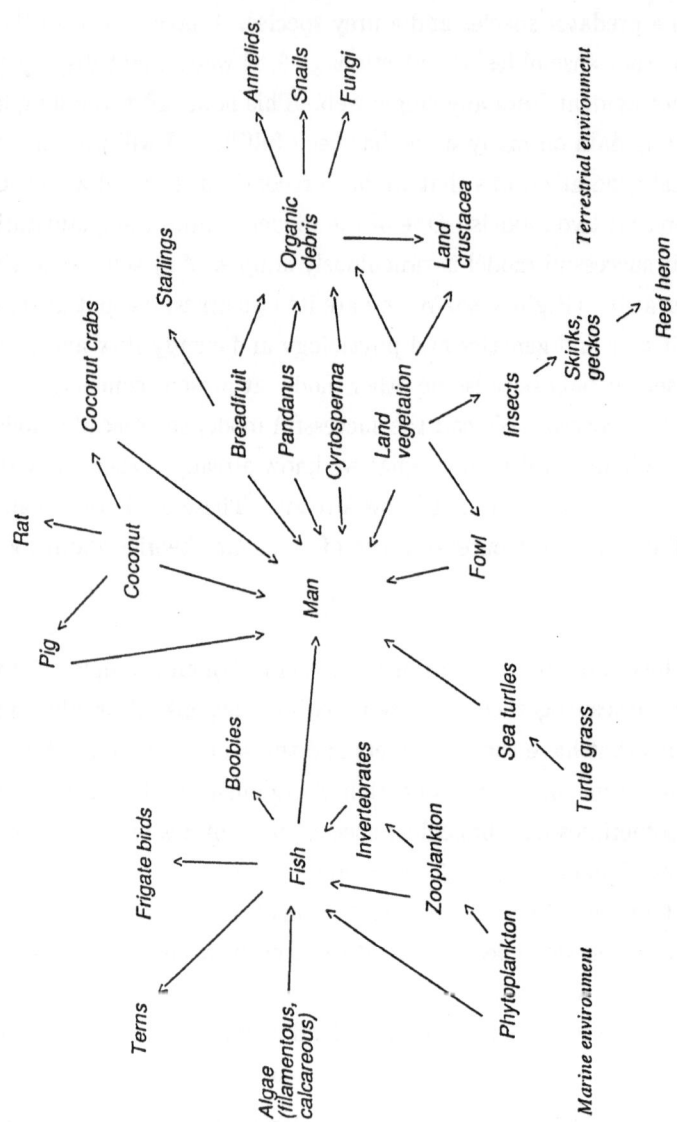

A chain is a path of links from a basal species to a top species, e.g., phytoplankton to fish to terns. The length of a chain is the number of links in it. In Fig. 1, the longest chain has only four links, and there is only one chain of length four. Short chains are typical of webs.

A cycle is a directed sequence of one or more links starting from, and ending at, the same species. A cycle of length 1 describes cannibalism, in which a species eats itself. Cannibalism is common in nature. But ecologists report cannibalism so unreliably that we have simply suppressed it from all the data even where it is reported. A cycle of length 2 means that A eats B and B eats A. In this example, as in most webs, there are no cycles of length 2 or more.

In summary, the terms just defined are trophic species, including top, intermediate and basal; links, including basal–intermediate, basal–top, intermediate–intermediate and intermediate–top; and chains, length (the number of links) and cycles.

## III.   LAWS

Here are five laws or empirical generalizations about food webs.

First, excluding cannibalism, cycles are rare. This generalization, without detailed supporting data, was offered as long ago as 1972 (Gallopin, 1972). Of 113 webs, three webs each contained a single cycle of length 2, and there were no other cycles (Cohen and Newman, 1985, p. 426; Cohen, Briand and Newman, 1986, p. 333).

Second, chains are short (Hutchinson, 1959). If one finds the maximum chain length within each web, then the median of this maximum in the 113 webs in the collection studied by Cohen, Briand, and Newman (1986) is four links and the upper quartile of the maximum chain length is five links. The longest chains in all 113 webs had ten links, and only one web had chains that long.

The last three laws deal with scale invariance (Cohen, 1977; Briand and Cohen, 1984; Cohen and Briand, 1984). We have compared the form of webs of different sizes. Such a comparison might be called the allometry of food webs. To appreciate the significance of what we found, consider a baby's face. The location of the eyes with respect to the top and the bottom of the head differs from the location of the eyes in an adult's face. That means that as the size of the face increases, the proportions change. I'm going to describe three laws which report that food webs, unlike a baby's face, have the same shape at different sizes. Scale invariance means that food webs of different size have constant proportions.

Our third law is scale invariance in three ratios: numbers of top species in proportion to numbers of all species; numbers of intermediate species in proportion to numbers of all species; and numbers of basal species in proportion to numbers of all species. Fig. 2. shows the proportions of all species that are top species, intermediate species and basal species. There's evidently no increasing or decreasing trend as the number of species increases (Briand and Cohen, 1984). The variability of proportions with respect to the average could be explained by chance alone. One can summarize crudely by saying that about a quarter (29 percent) of all species are top, about half (53 percent) are intermediate and about a quarter (19 percent) of the species are basal. Here, scale invariance describes the observation that as the number of species in 62 webs varies from 0 to 33, the proportions of top, intermediate and basal species apparently remain invariant.

Our fourth law is scale invariance in the proportions of the different kinds of links. In Fig. 3a (Cohen and Briand, 1984), for example, the abscissa is the number of species and the ordinate is the proportion of basal–intermediate links among all links. There is no clear evidence of an increasing or decreasing trend. The proportions of different kinds of links, like the proportions of species, are approximately scale–invariant. Here the scatter about a horizontal line is too big to be explained by random sampling.

The fifth law is that the ratio of links to species is scale–invariant. This turns out to be fundamental. Fig. 4 plots the observed number of links in each food web against the observed number of species, for 113 webs (Cohen, Briand and Newman, 1986). We find a straight line with slope about 2. That means that a web of 25 species has on average about 50 links. We first came across this generalization with 62 webs (Cohen and Briand, 1984). Then Briand collected an additional 51 webs, and we found (Cohen, Briand and Newman, 1986) that the new data superimpose beautifully on the old data. This scale–invariant ratio of links to species is a consistent feature of nature, not something we have invented.

In summary, I have reviewed evidence for five "laws" of food webs. Qualitatively, these laws state that cycles are rare, chains are short, and there is scale–invariance in the proportions of different kinds of species, in the proportions of different kinds of links, and in the ratio of links to species. We have quantified each of these laws.

4.     Models

Let me turn now from empirical regularities to models. Here some mathematics is inevitable. This reminds me of a story.

A mathematician and an ecologist were sharing a cell the night before their execution (for crimes unimaginable). The executioner came in to ask their last wishes.

The mathematician looked over at the ecologist and said, "I've been doing some work in mathematical ecology. I have some interesting results. Before I die, I would like to give a seminar on my work to an ecologist."

"Certainly", said the executioner, "we'll arrange it tomorrow morning." He then turned to the ecologist. "And what would you like?"

The ecologist said, "I would like to be executed before the seminar."

Let  S  denote the number of trophic species and  L  the number of links. We enumerate all the species along both the rows and columns of a "predation matrix," a square table of numbers with  S  rows and  S  columns. Name the matrix A. We put a  1  in the intersection of row  i  and column  j  if the species labeled  j  eats the species labeled  i, and a  0  otherwise. Since I am excluding cannibalism, all the diagonal elements (where i = j) are  0.

Fig. 2 Three ratios, plotted as a function of the number of species, show scale–invariance in the proportions of species. The fitted lines are constrained to be horizontal. (a) Top species/total species. The height of the line is 0.2853. (b) Intermediate species/total species. The height of the line is 0.5251. (c) Basal species/total species. The height of the line is 0.1896. From p. 265 of Briand and Cohen, 1984.

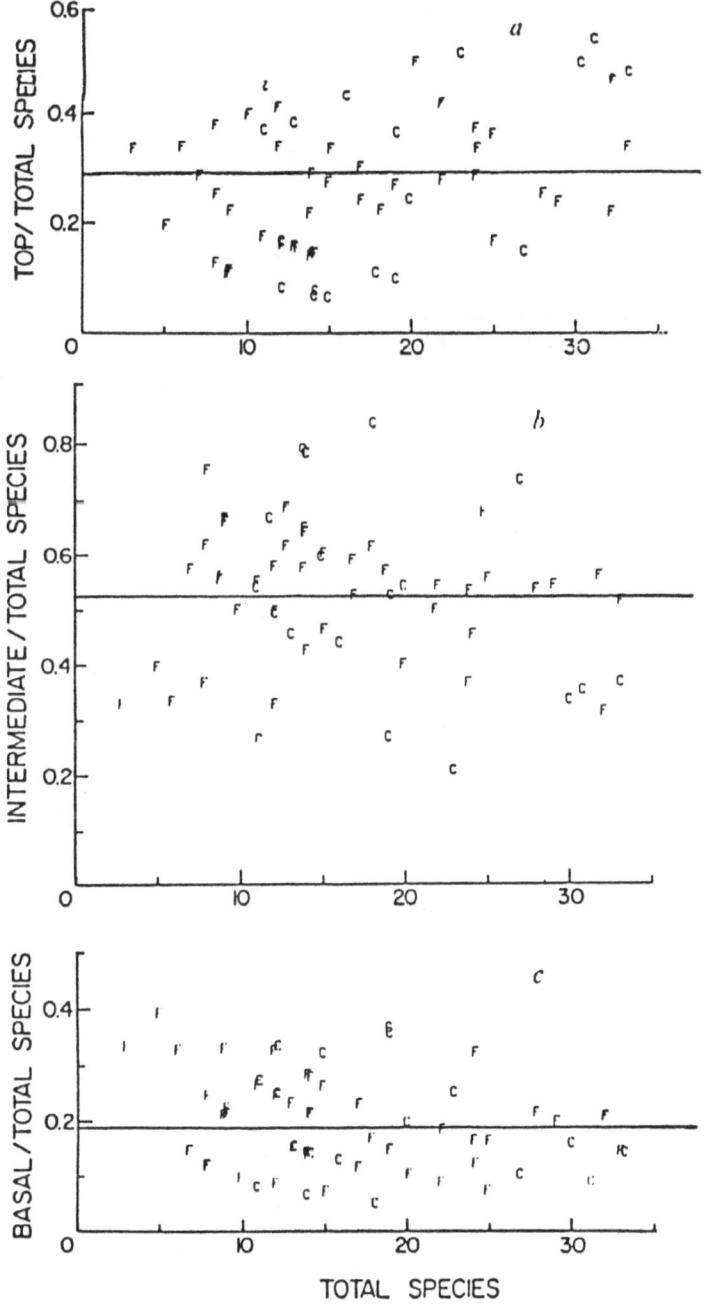

Fig. 3. Four ratios, plotted as a function of the number of species, show scale–invariance in the proportions of links. The fitted lines are constrained to be horizontal. (a) Basal–intermediate links/total links. The height of the line is 0.274. (b) Basal–top links/total links. The height of the line is 0.077. (c) Intermediate–intermediate links/total links. The height of the line is 0.301. (d) Intermediate–top links/total links. The height of the line is 0.348. The points in the upper left corner of (a) are based on very few links. From p. 4107 of Cohen and Briand, 1984.

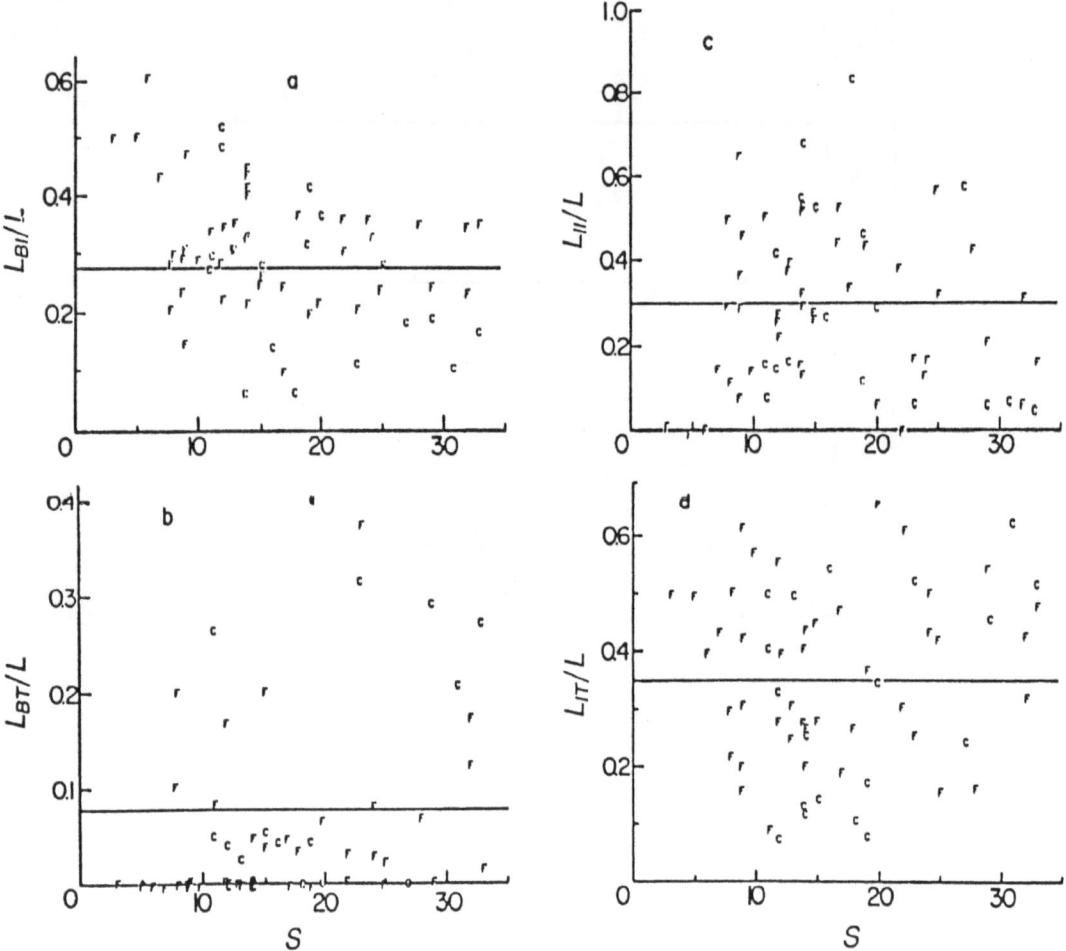

Fig. 4. Observed number L' of links as a function of the observed number S' of species in 113 webs. From p. 335 of Cohen, Briand and Newman, 1986.

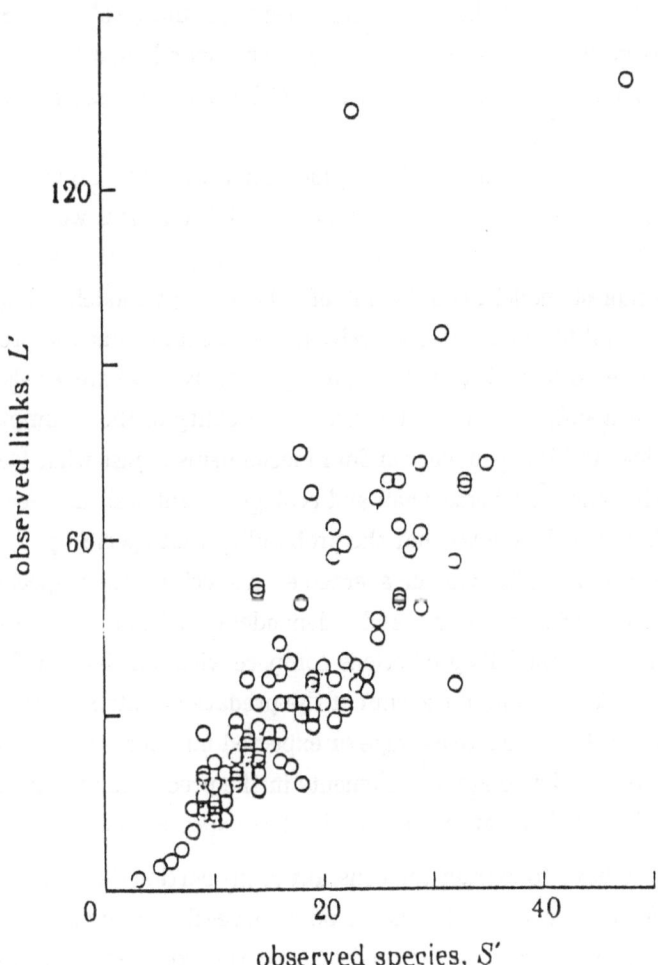

observed links. $L'$

120

60

0

20

40

observed species, $S'$

In terms of this predation matrix, the total number of links is the sum of the elements of A. The sum picks up a  1  if there is a link from prey to predator and a  0  if there is no link.

The predation matrix also tells whether a species is top. If a species is top, then nobody eats it. That means that the row of that species should be all  0's. So a 0–row corresponds to a top species. Similarly, a 0–column corresponds to a basal species because the species is not eating anything. A species that has neither a 0–row  nor a 0–column is intermediate.

I am going to present first a model that does not work. The calculation in this model is simple and gives the flavor of a more complicated model that does work. For the model that does work, I will just describe the results without going through the calculations.

Here is the simplest model I could think of: the anarchy model. I hope you will agree that it has some beautiful features. The anarchy model assumes that the probability that any species  j  eats any other species  i  is just c/S, independently of whatever else is going on in the food web. That is a simple model. (The brazen unreality of the assumption that all species act by identical and independent random mechanisms is just what lends verisimilitude to the story about the jailed mathematician and ecologist. But wait and see what emerges from this tissue of fiction!) It follows that the probability that species  j  does not eat species  i  is $1 - c/S$. On the average each species eats  c  species chosen from among the  S  possible species, randomly and independently of all other species.

How do the anarchy model's predictions compare with our five laws? The expected number of links is the expectation of the sum of the predation matrix elements. As is conventional, I will use  E  to denote average or expected number, so  E(L)  denotes the expected number of links. There are  $S^2$  elements in the predation matrix A and the probability is  c/S  that an element  $a_{ij}$  equals 1. The expected sum of the elements is  $S^2 \times$

$c/S = cS = E(L)$. We have to extract the constant of proportionality, i.e., the slope in Figure 4, from the data. We take  c = 2. That is the only curve–fitting in this model. Everything else is derived. Thus, the anarchy model predicts that the expected number of links should be proportional to the number of species, as observed. The links–species scaling law fits quantitatively because we made it fit by taking  c = 2.

Now I show that the anarchy model predicts qualitatively the scale–invariance in the proportion of top species, but gets the proportion wrong quantitatively. Since there are  S  species, the expected number of top species is  S  times the probability that any one species is a top species. The probability that one species is a top species is the probability that the sum of elements in some row is  0. The row sum is  0  if and only if every elements is  0. So  E(T) is  S  times the probability, raised to the  S  power, that each row element is zero. The probability that a single element is  0  is  $1 - c/S$. As  S  gets big, $(1 - c/S)^S$  approaches  $e^{-c}$. So  E(T)/S, the expected fraction of top species, is asymptotically (for large S)  $e^{-c}$. Qualitatively, this is good. It means that as the number of species increases the fraction of top species does not change (more accurately, the fraction approaches a limit). This simple model predicts scale–invariance of species proportions.

However, if $c = 2$ in this formula, $e^{-c} \approx 14$ percent. The species scaling law is qualitatively good but quantitatively poor because 14 percent is too small—it is not near the one–quarter (let alone 29 percent) shown in Figure 2.

What else does this model predict? The probability that a web has at least one 2–cycle as S goes to infinity is $1 - e^{-c^2/2}$. If $c = 2$, the model predicts that 86 percent of webs should have one or more 2–cycles. That is not good because we found 2–cycles in only 3 of 113 webs.

The principal problems with the anarchy model are that it predicts too many cycles and that it fails to predict the proportion of top species. Let's fix one problem at a time and see whether that solves any other problems as well. We get rid of the problem with cycles by fiat in the next model, the cascade model.

I am now going to describe the cascade model, but not the calculations required to squeeze results out of it. Assume S species. Somehow nature numbers them from 1 to S (without showing us the numbering). Any species $j$ in this hierarchy or cascade can feed on any species $i$ with a lower number $i < j$ (which does not mean that $j$ does feed on $i$, only that $j$ can feed on $i$). However, species $j$ cannot feed on any species with a number $k$ at least as large, $k \geq j$. The cascade model assumes that each species actually eats any species below it with some probability $d/S$, independently of whatever else is going on in the web. (I have changed notation for the probability parameter from $c$ to $d$ so as not to mix up the anarchy and cascade models.)

In the predation matrix A, $a_{ij}$ is 0 always if $i \geq j$. The predation matrix with this labelling is strictly upper triangular. An element above the diagonal $(i < j)$ is 1 with probability $d/S$ and is 0 with probability $1 - d/S$, and all elements are independent.

To derive predictions from the cascade model, we must take one number from nature. To simplify slightly, we estimate $d$ approximately as twice the observed number of links divided by the observed number of species; we multiply the number of links by two here because roughly half of the matrix is empty. As the number of species becomes large, the cascade model predicts 26 percent top species, 48 percent intermediate species and 26 percent basal species. We observed 29 percent top species, 53 percent intermediate and 19 percent basal. We predict the following percentages of basal–intermediate, basal–top, intermediate–intermediate and intermediate–top links: 27, 13, 33 and 27. We observed, correspondingly, 27, 8, 30 and 35.

I think it is nice that the cascade model reproduces all the laws of scale–invariance qualitatively, but far more striking that the cascade model gives a remarkable quantitative agreement between observed and predicted proportions. We put one number $d$ into the cascade model and get out five independent numbers (because the three species proportions have to add up to 1 and the four link proportions have to add up to 1). I would like to emphasize that these predictions use only the observed ratio of links to species.

For a finite number of species, we calculated from the cascade model the expected fraction of top species and the predicted variance. Figure 5 shows that the cascade model predicts not only the means but also the variability in the proportion of top species. We do not know whether the cascade model can predict the variability in proportions of links because we do not know how to calculate analytically what variability the cascade model predicts.

The cascade model was built to, and does, explain qualitatively and quantitatively the mean proportions of different kinds of species and links. Can the cascade model describe the number of chains of each length counting all the possible routes from a basal species to a top species?

Let me illustrate with an artificial example (Figure 6) how to get a frequency histogram of chain length from a food web. The link from 1 to 2 is a chain of length 1. The path 1, 3, 4 is a chain of length 2, and the path 1, 3, 5 is another chain of length 2. A numerical summary of the chain length distribution of the web in Figure 6 is that it has one chain of length 1, two chains of length 2 and no longer chains.

Figure 7 shows the expected number of chains of each length, according to the cascade model, using parameters of a typical web, namely 17 species and d close to 4. Figure 7 also shows the results of one hundred computer simulations of the model using the same parameters. The sample mean numbers of chains of each length agree well with theoretically expected number calculated from the model. That agreement increases the probability that both the calculations and the simulations are right.

To see how well the cascade model predicts the observed distribution of chain length of a given <u>real</u> web, we generated random webs according to the cascade model with the parameters of the observed web. We measured how often the chain length distribution of a random web was further from the chain length distribution predicted by the cascade model than the real observed chain length distribution was from the predicted distribution. We used two measures of goodness of fit: the sum of squares of differences and a measure like Pearson's chi–squared. If the discrepancy between the observed and the expected frequency distributions was smaller than most of the discrepancies between webs randomly generated according to the cascade model and the mean frequency distribution expected from the model, we said the fit was good. If the discrepancy between observed and
simulation is 0.003. From p. 324 of Cohen, Briand, and Newman, 1986.
predicted chain length distributions was bigger than most simulated discrepancies, we said the fit was bad.

Have no illusions about what a good fit means. Food web 18 in Figure 8 illustrates a good fit. Food web 53 illustrates a poor fit.

Of 62 webs in Briand's original collection, 11 or 12 (depending on the measure of

Fig. 5. The predicted mean proportion of top species (middle line) and a confidence interval of $\pm 2$ standard deviations (upper and lower lines) as a function of total species S, according to the cascade model. X is constant environment, o is fluctuating environment. The symbols X and o have been perturbed from their exact locations by a small random amount to indicate when several food webs have exactly the same coordinate. The data are replotted from Briand and Cohen, 1984. From p. 436 of Cohen and Newman, 1985.

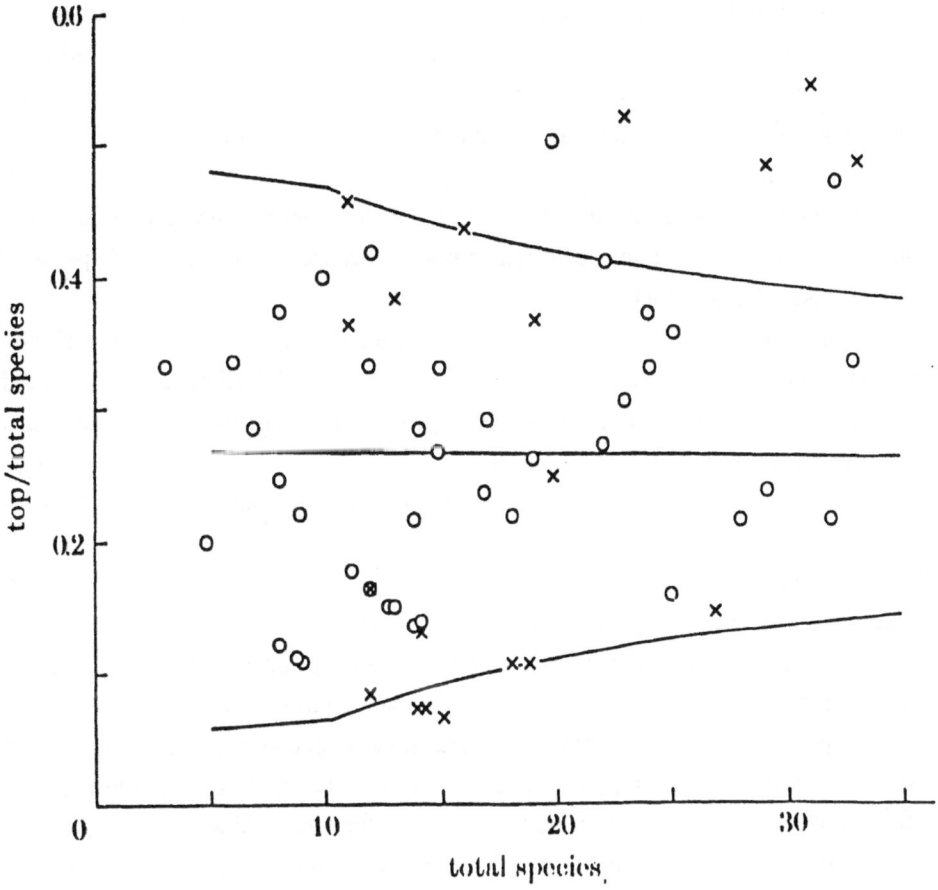

goodness of fit used) were badly described by the cascade model. The model's success with 50 or 51 of these webs made us afraid that we had overfitted the model to the data. Perhaps, by constructing the cascade model to explain the mean proportions of top, intermediate and basal species and the proportions of different kinds of links, we had used so much information from the data that there was no possibility for the fits to the chain length distribution to be bad, even though they were not used to build the model. This worried us. So Frédéric Briand found and edited 51 additional webs which we had never analyzed before. The ratio of links to species was the same for these new webs as for the old webs, as I mentioned already. With these fresh data, we found only five webs with poor fits to the cascade model's predicted frequency distribution of chain length. The proportion of poor fits, 5 of 51 webs, was smaller among the new webs than it had been among the original webs. This gave us some confidence (in addition to considerable surprise and pleasure).

The cascade model uses no information about chain length to predict the frequency distributions of chain length! The predictions derive solely from the number of species and the number of links. No parameters are free.

Let me moderate this final burst of enthusiasm for the empirical successes of the cascade model by emphasizing that the model needs to be tested further, tested until it fails, as it surely will. How well can the cascade model predict the moments of chain length (as Stuart Pimm has asked), or patterns of omnivory and intervality (Cohen, Briand and Newman, 1986)? Not all the evidence is in yet. Many good questions remain to be asked.

## V.    PREDICTIONS

What new predictions does the cascade model make?

One prediction may not be new, but receives a new foundation from the cascade model. This prediction is that, with improved data, the number of basal species should equal the number of top species. The cascade model predicts equal expected numbers of top and basal species because the model has complete symmetry between top and basal species. This prediction supports what Stuart Pimm had been saying for years. His suspicion (1982) that ecologists are more interested in the feathered and furry animals at the top of the web than in the slimy and creepy animals at the bottom is consistent with everything that we have found.

Secondly, in large webs ($S > 17$), the cascade model implies a rule of thumb which I have never seen stated in the ecological literature: The mean length of a chain should equal the mean number of prey species plus the mean number of predators of an average species. Both should equal a number near 4. That rule of thumb follows from two separate analyses of the cascade model.

Can the cascade model explain qualitatively why the longest chains in webs are typically short? Figure 9 shows the relative expected frequency of various chain lengths as the number of species goes to infinity, according to the cascade model. Practically, no chains have length 8, 9 or 10. A beautiful piece of mathematics due to Charles M. Newman shows

Fig. 6 Hypothetical food web to illustrate how the frequency distribution of chain lengths is counted. There is one chain of length 1 (from species 1 to species 2) and there are two chains of length 2 (from species 1 to species 4 and from species 1 to species 5).

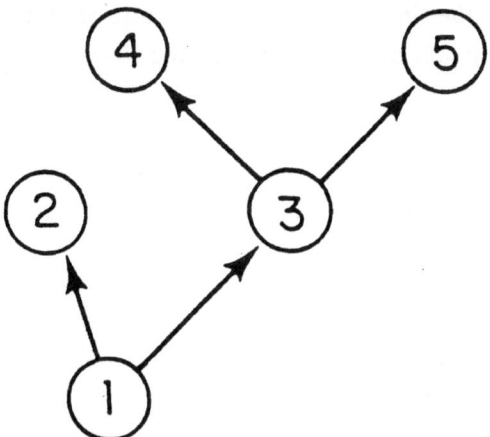

Fig. 7. Theoretically expected number (solid line) of chains of length 1 to 9 in a web of
$S = 17$ species, according to the cascade model with $c = 3.75$, sample mean number (0) of
chains of each length in 100 simulations of the cascade model, and sample mean plus one
sample standard deviation (□) in the number of chains of each length. No chains with more
than nine links occurred in the simulations; the expected total number of such chains per
simulation is 0.003. From p. 324 of Cohen, Briand, and Newman, 1986.

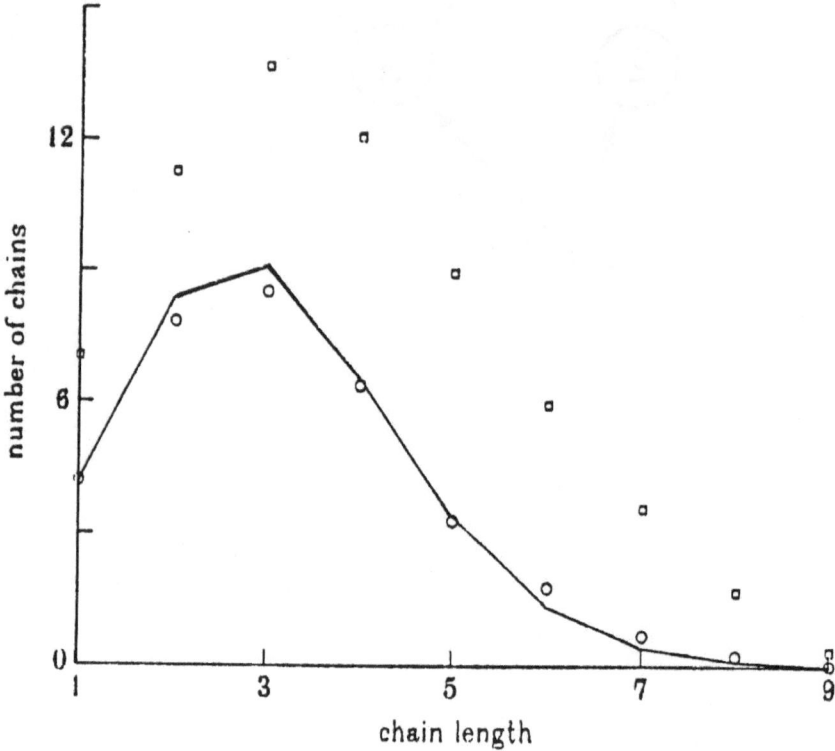

that, in very large webs, the longest chain grows like log S/log log S. That is very slow growth. In a web with $10^{18}$ species, which is probably an upper bound for the world, the cascade model predicts that the longest chain will almost never have more than 20 links.

## VI.    APPLICATIONS

What good is all this for the real world of practical affairs? Let me speculate about four ways this work may contribute to human well–being.

First, environmental toxins cumulate along food chains. An understanding of the distribution of the length of food chains is essential, though not sufficient, for understanding how toxins are concentrated by living organisms.

Second, an understanding of the invariant properties of food webs is essential for anticipating the consequences of species' removals and introductions. Such perturbations of natural ecosystems are being practiced with increasing frequency in programs of biological control. So far, people have not been very successful at anticipating all the consequences of introducing or eliminating species. An understanding of food webs should help anticipate the consequences.

Third, an understanding of food webs will help in the design of nature reserves and of those future, mobile nature reserves that will be required for long–term manned spaceflight. A nature reserve with all top species would be expected to have trouble, according to the cascade model. For humans to survive and to be fed in space, we need to know more about the care and feeding of food webs.

Fourth, and finally, since food webs include man, perhaps an understanding of such webs will give us a better understanding of man's place in nature, here on earth. It is a remarkable fact that we have not detected any consistent pattern of difference between those webs in which man is a species and those webs in which man is not a species.

Of course, as a graduate student pointed out to me when I said this, we have not looked yet at agricultural ecosystems strongly influenced by man. When we look at a new class of food webs, we might see new patterns. God created graduate students to keep us all honest.

Fig. 8. Examples of "acceptable" and "poor" fits between the predicted (mean) numbers of chains of each length according to the cascade model and the observed numbers of chains of each length. In the serial numbering of Briand (1983), which is used here, number 18 is the Kapingamarangi Atoll food web (see Fig. 1 above) of Niering (1963) and number 37 is the California sublittoral (sand bottom) food web of Clarke et al. (1967). These webs correspond (see Briand, 1983) respectively to food webs numbered 11 and 2 by Cohen (1978), who gives the predation matrices in full. For food web 18, four chains of length 4 are shown while Fig. 1 has one chain of 4 links. The reason for this discrepancy is that Cohen (1978) added to the predation matrix for this web links that Niering (1963) described in his text but omitted from his figure.

| | food web 18 "acceptable" fit | | | food web 53 "poor" fit | |
|---|---|---|---|---|---|
| chain length | predicted frequency | observed frequency | chain length | predicted frequency | observed frequency |
| 1 | 6.5 | 13 | 1 | 6.5 | 1 |
| 2 | 9.9 | 10 | 2 | 8.6 | 19 |
| 3 | 8.6 | 5 | $\geq 3$ | 12.4 | 0 |
| 4 | 5.2 | 4 | | | |
| $\geq 5$ | 3.6 | 0 | | | |

Fig. 9. Asymptotic relative expected frequency of chains of each length, in webs with an arbitrarily large number of species, according to the cascade model with c = 3.71. Modified from p. 361 of Newman and Cohen, 1986.

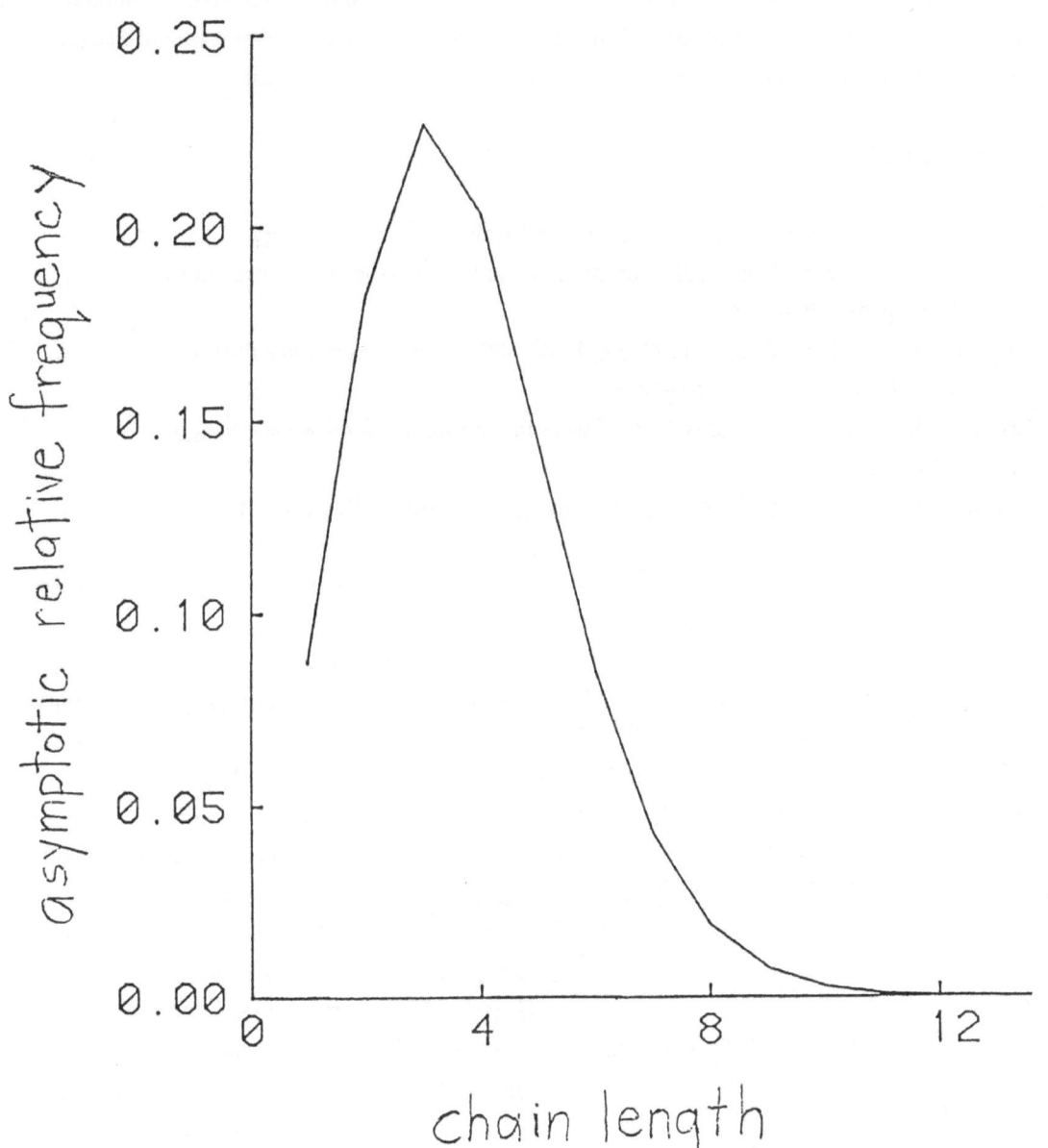

## Acknowledgements

This work was supported in part by U.S. National Science Foundation grant BSR 4–07461, a Fellowship to J.E.C. from the John D. and Catherine T. MacArthur Foundation, and the hospitality of Mr. and Mrs. William T. Golden. Charles M. Newman and Stuart Pimm made numerous corrections and improvements to a previous draft.

## REFERENCES

Briand, F. 1983 Environmental control of food web structure. Ecology 64, 253–263.

Briand, F. and Cohen, J. E. 1984 Community food webs have scale–invariant structure, Nature 307, 264–266.

Clarke, T.A., Flechsig, A.O., and Grigg, R.W. 1967 Ecological studies during Project Sea Lab II. Science 157, 1381–1389.

Cohen, J. E. 1977 Ratio of prey to predators in community food webs. Nature 270, 165–167.

Cohen, J. E. 1978 Food Webs and Niche Space. Princeton: Princeton University Press, 189 pp.

Cohen, J. E. and Briand, F. 1984 Trophic links of community food webs. Proc. Nat. Acad. Sci. U.S.A. 81, 4105–4109.

Cohen, J. E., Briand, F., and Newman, C. M. 1986 A stochastic theory of community food webs. III. Predicted and observed lengths of food chains. Proc. Roy. Soc. (London) B 228, 317–353.

Cohen, J. E., and Newman, C. M. 1985 A stochastic theory of community food webs. I. Models and aggregated data. Proc. Roy. Soc. (London) B 224, 421–448.

Cohen, J. E., Newman, C. M. and Briand, F. 1985 A stochastic theory of community food webs. I. Models and aggregated data. Proc. Roy. Soc. (London) B 224, 421–448.

Cohen, J. E., Newman, C. M. and Briand, F. 1985 A stochastic theory of community food webs. II. Individual webs. Proc. Roy. Soc. (London) B 224, 449–461.

DeAngelis, D. L., Post, W. M. and Sugihara, G. (eds.) 1983 Current Trends in Food Web Theory. ORNL–5983. Oak Ridge, Tennessee: Oak Ridge National Laboratory.

Diamond, J. M. and Case, T. J. (eds.) 1985 Community Ecology. Cambridge: Harper and Row. 665 pp.

Hutchinson, G. E. 1959 Homage to Santa Rosalia or why are there so many kinds of animals? American Naturalist 93, 145–159.

Kikkawa, J. and Anderson, D.J. (eds.) 1986 Community Ecology: Pattern and Process. Melbourne and Oxford: Blackwell Scientific. 432 pp.

Newman, C. M. and Cohen, J. E. 1986 A stochastic theory of community food webs. IV. Theory of food chain lengths in large webs. Proc. Roy. Soc. (London) B 228, 355–377.

Niering, W. A. 1963 Terrestrial ecology of Kapingamarangi Atoll, Caroline Islands. Ecological Monographs 33, 131–160.

Pimm, S. 1982 Food Webs. London: Chapman and Hall. 219 pp.

Pimm, S.L. In press, The geometry of niches. This volume, Chapter 7.

Sugihara, G. 1984 Graph theory, homology, and food webs. Proc. Symp. Applied Math. 30, 83–101. Providence, RI: American Mathematical Society.

# CHAPTER 7
## The Geometry of Niches

Stuart L. Pimm
Department of Zoology and
Graduate Program in Ecology
The University of Tennessee
Knoxville, TN 37996 U.S.A.

## I.    INTRODUCTION

Like others in this volume, this chapter has a lot to do with scale. But it is organizational scale rather than spatial or temporal scale that I wish to consider. Much of what is commonly called community ecology deals with a very small number of species — often two. Yet communities are large sets of species — all the species in an area or at least those in a particular taxonomic or trophic group. Clearly, we need to ask how species interactions are built into large sets. Such is the aim of the various studies of food webs: food webs are the diagrams depicting which species in a community interact trophically. Among the achievements of recent food web studies is a long catalogue of food web patterns reviewed recently by Lawton (1989). These patterns can be grouped into six broad categories: the general patterns of connectance and trophic grouping across trophic levels (e.g. Briand and Cohen, 1984, Cohen, this volume, Cohen and Briand, 1984, Yodzis, 1981), the number of trophic levels (e.g. Cohen, this volume, Pimm and Lawton, 1977, Pimm and Kitching, 1987), the patterns of species which feed across trophic levels (e.g. Pimm and Lawton, 1978), the degree to which interactions are grouped into compartments (e.g. Pimm and Lawton, 1980, Yodzis, 1982), and such ratios as that between the number of species at one trophic level and the number of species at the trophic level on which these predators feed (e.g. Cohen, 1977).

The sixth category involves the patterns of which predators feed on which prey species. The most basic question we must ask when we go from considering whether two predators potentially compete (by virtue of their sharing prey species) or from considering a one–predator one–prey model, to our considering interactions at the community scale, is how these interactions are organized. It is patterns in this sixth category that I wish to consider in this paper.

Two predators either overlap in their choice of prey, or they do not. While this is hardly interesting geometrically, considering large sets of predators raises the possibility that certain topological patterns, well–known in mathematics, may predominate in food webs. The first attempt along these lines was Cohen's studies of intervality (Cohen, 1978). More recently, Sugihara (1983, 1984) has analyzed a suit of characters including asteroids, holes and rigid circuits, which presently I shall define and consider.

Figure 1.(a) An interval food web and (b) a non–interval web. Lines below the food webs
show the patterns of overlaps among the four predatory species. Most food webs are
interval. From Pimm (1982).

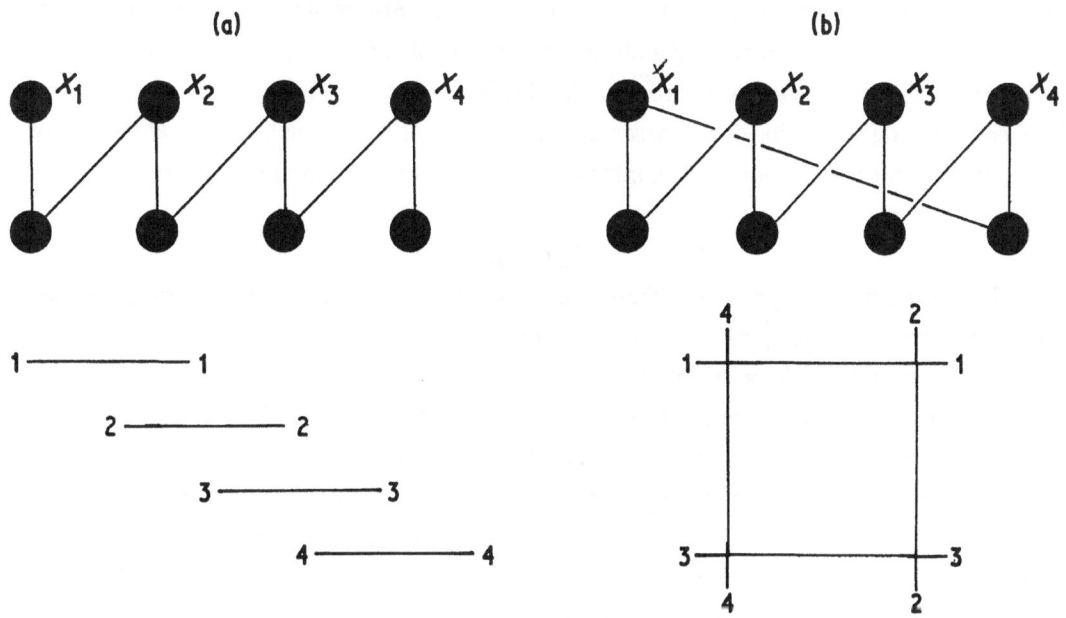

## Intervality

A food web is deemed interval if the overlaps in the predators' uses of prey species can
be expressed as possibly overlapping segments of a line. Fig. 1a shows a simple interval food
web and the line segments that describe the patterns of overlap. Fig. 1b shows an only
slightly more complex web. The overlaps cannot be expressed along a line, though they can
be expressed in two dimensions. This web is not interval.

Cohen (1978) showed that food webs are more interval than chance alone dictates.
Cohen used computer simulations to generate the expected probability of a web being interval
under each of seven models. In each of the seven models, species interactions were chosen, at
random, subject to model–specific constraints, designed to ensure that the interactions were
ecologically sensible. Now, whether we expect a model food web to be interval or not depends
a good deal on the number of predators in the web. Non–intervality is only possible with four
or more predators. And, for four predators, it is a sensitive result: the non–interval pattern of
Fig. 1b is destroyed if more interactions are added or if the interactions are placed differently.
For the model Cohen considered to be best, a graph showing the probability that a web will
be interval versus the number of predators, shows a sharp transition (Pimm, 1982). Below 12
predators both model and real webs were always interval. Above 12 predators, the models
and most of the real webs were non–interval. There were, however, a few, large interval real
webs and it is these that lead Cohen to his conclusion on the predominance of interval webs in
nature.

An obvious problem is that factors which tend to group predators into distinct sets of small numbers of species (i.e. <12), will greatly enhance the chances of intervality. Grouping by trophic levels or into compartments, for example, would increase the chances of intervality in a web with a large number of species. As Critchlow and Stearns (1982) point out, Cohen's models do a poor job of predicting the observed numbers of such groupings. So, is Cohen's conclusion correct? The answer is provided by Sugihara (1982), who examined the groupings of Cohen's original webs plus others assembled by Briand (1983). Sugihara was able to show that, even within these groupings, the interval pattern was more common that we would expect by chance.

## Rigid circuits

Drawing out which predators share prey produces the predator overlap graph. These graphs show another topological regularity demonstrated by Fig. 2a. Consider an approximate physical analogy, where we make a physical model of these overlaps by connecting predators that share one or more prey. The physical model of Fig. 2a would be rigid, not floppy, because when there are connections between four or more points (predators) these connections (overlaps) are triangulated. (Note that this is not the case for the web in Fig. 2c., and that species 8 in Fig. 2a does not violate the condition, because it is not part of a closed circuit of four or more species.) Technically, this property is called a rigid circuit graph and real food webs contain an overwhelming preponderance of them (Sugihara, 1982, 1983, 1984).

The rigid circuit property does not ensure that food webs are interval. It is possible to draw a non-interval, rigid circuit predator overlap graph: Fig. 2b is an example. The graph is considered to be rigid circuit, because there are no circuits around four or more points. For obvious reasons, this pattern is called 'asteroidal'. As Sugihara mentions, if a web is not asteroidal and it is rigid circuit, then it will be interval.

## Holes

Finally, let us go from looking at the predators' views of their prey, to the preys' views of their predators, to form what I call assembled prey overlap graphs. We shall see that these graphs synthesize a good deal of information. Suppose we form solids (technically, simplexes) by connecting prey species that share a particular predator. Two such prey would form a line, three prey would form a triangular plane; four a tetrahedron, and so on. Planes are easier to draw than multisided solids and I shall select my examples accordingly. Some prey will be used by more than one predator: these will link together the different individual solids. Fig. 3a shows a combination of solids for three predators which each exploit 3 of a total of 6 prey species. Fig. 3b is the corresponding food web. Pursuing the physical analogy, we observe there is a hole between prey species 2, 3, and 5. These structural holes are exceedingly rare in

Figure 2. Predator overlap graphs connect predators that share one or more prey species. Both (a) and (b) are rigid circuit, (c) is not. The predator overlap graph for (a) has an interval representation, (b) is asteroidal (it has three or more 'tentacles', from a central species (or, in this case a group of species). The predator overlap for (c) is not interval. Key (a) Bird's (1930) study of a Canadian willow forest. 1, a fungus, 2, insects, 3, another group of insects, 4, three species of birds, 5, another three species of bird, 6, spiders, 7, a frog, 8, garter snake. (b) Kohn's (1957) study of predatory gastropods of the genus _Conus_ on the subtidal reefs of Hawaii. (b) All species are of the genus _Conus_: 3, _ebraeus_, 4, _chaldeus_, 5, _miles_, 6, _rattus_, 7, _distans_, 8, _vexillum_, 9, _vitulinus_, 10, _imperialis_. See also Table 1. (c) This pattern is so rare that I have no natural example: this graph is just hypothetical.

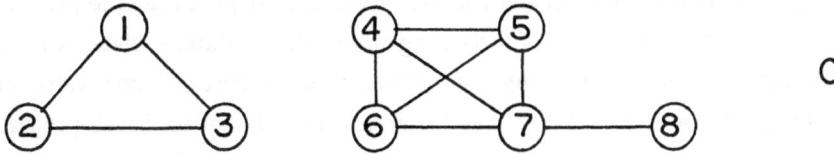

a

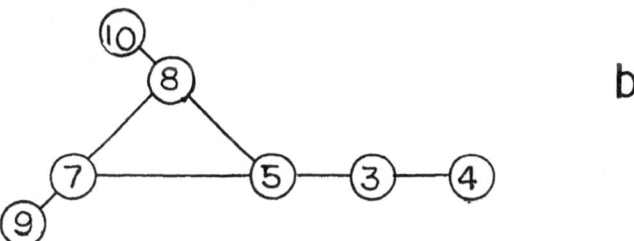

b

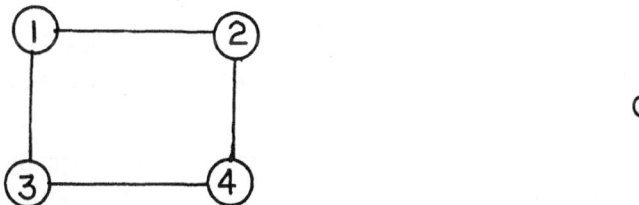

c

real food webs (Sugihara, 1982, 1983, 1984). In contrast, consider the system in Fig. 3c and d; it resembles that in 3a and 3b, but predator C now feeds on prey species 3 and so there is no hole.

The lack of holes implies certain rules for community assembly or dis–assembly (Sugihara, 1982). For example, if we start to demolish the trophic structure of the community, certain interactions cannot disappear first. The interaction between predator C and prey 3 cannot be removed before those between C and 2 and 5, or else a hole is created.

Why do these regularities occur and why are some alternatives (like the existence of holes) so very rare? Consider Fig. 3e and f. We have removed the hole by removing the interaction between A and 3 and, for convenience, drawn 3 over to the right of the web. We see that the prey now have a common ranking among the predators. This could be by size, or by food quality, or indeed by any other variable. We could imagine that the prey, in the sequence 1,2,4,5,6,3, represent items of increasing size in the predators' diets. Critchlow and Stearns (1982) call this common ranking the 'interval diet' property of food webs. (I find this a little confusing, because 'interval diet' and 'interval' are different food web properties — the latter referring to overlaps. I shall use 'prey ranking' instead.) Reversing the argument, we see that when predators do rank prey, holes are destroyed. This is not even a necessary condition for the absence of holes. In Fig. 3c and d, the predators do not rank the prey (4 is out of sequence), yet there is no hole. In this web, if the prey species were to be ranked in increasing order of size, we would wonder why species 4 is absent from the diet of predator B — which takes prey smaller (3) and larger (5) than species 4.

Simple feeding constraints will make it unlikely that while some predators will take, say, small and medium prey, and others medium and large, that others will take only small and large. Such feeding constraints imply ranking of the prey and a reduction of the chances of getting holes in the resource graph. Ranking also precludes non–interval food webs.

How much of the rarity of holes or of non–interval food webs is simply due to the predators ranking the prey? In only about 50% of the set of predators examined by Critchlow and Stearns can the prey be ranked. So the rarity of holes and interval overlaps is not totally explained by the predators choosing a simple ranking of their prey species.

II. THE IMPACT OF DIETARY OPPORTUNISM

Tabulations of the use of prey, habitats, or other resources, by a set of predators are abundant in the ecological literature. Ecologists are practised at interpreting them: we all know that species tend to take different resources in different proportions. Indeed, the literature on such resource partitioning is voluminous (Schoener, 1974 is a review). None the less, resource partitioning is only a tendency away from a major feature of such tables: the common utilization of prey species by the various predators which, in these studies, often share not only habitat but taxonomic affinity.

In addition to this rather obvious feature, there is the equally obvious feature that some predators and prey are more common than others. Typically, the commonest prey

Figure 3. Parts (b), (d) and (f) are food webs: A,B, C are predators and 1–6 are their prey. Parts (a), (c), and (e) are graphs formed by connecting prey species which share a predator. Thus, in (a) 1,2 and 3 are connected because predator A feeds on all three species. Prey 3 is also connected to 6, because B feeds on both. But, 1 and 4 are not connected — they have no predator in common. Furthermore, prey for each predatory species form solids (planes in all cases except for the tetrahedron connecting 2,3,4 and 5 in (c) and the line connecting 1 and 2 in (e).) This recipe forms a hole between 2, 3 and 5 for the food web in (b), but the hole is absent from webs d and f. Holes are exceedingly rare in real food webs.

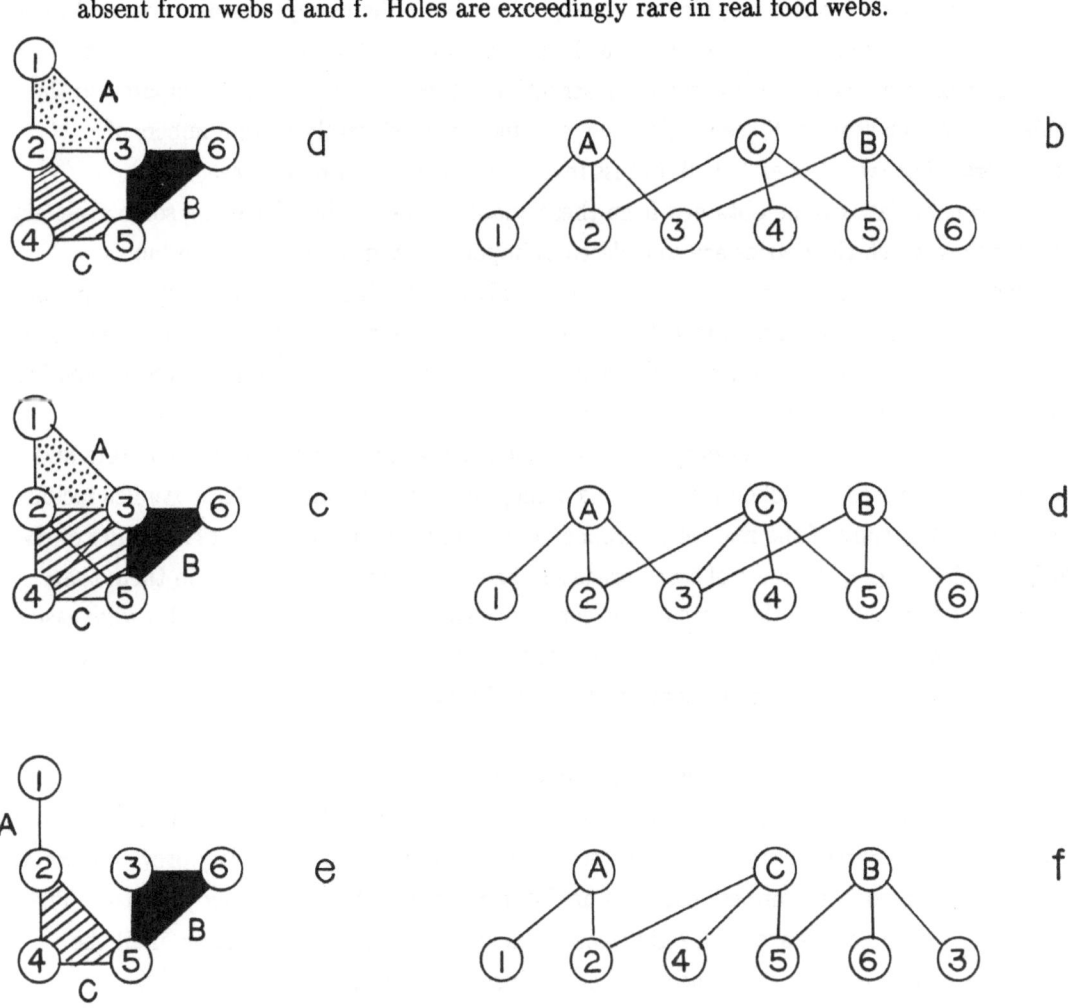

species in the diet of one predator species will be at least well–represented in the diet of another predator. It might even be the commonest prey in the second predator's diet. These features, combined with obvious sampling problems, are sufficient to explain the high incidence of webs that are rigid circuit, that are interval, and which lack holes. When we factor out the effects of such dietary similarities, the niche patterns are no longer so unexpected.

First, a comment about the data on which I shall base my arguments. Most analyses of food web patterns are based on binary data — either A eats B or else it does not. But, for the patterns involving two trophic levels — predators and their prey — quantitative data are sometimes available. Such data are often the number of prey found in the stomachs of predators or, more rarely, are based on observations of feeding predators. Compared to the bulk of food web studies, these studies may also include moderately large numbers of predators. The two studies I shall discuss involve about a dozen predators.

Binary data can be obtained from these quantitative studies by setting some arbitrary level, below which the entries are considered unimportant and so equated with 'no interaction'. It is to be hoped that studies that only report binary data use such a criterion implicitly for deciding which interactions between species to report. Quantitative studies are particularly valuable because, by varying the arbitrary level, we can see if the recognized food web patterns are sensitive to the chosen level.

As a starting point, we can take any of the six quantitative studies of predators and their prey discussed by Cohen (1978) in his monograph on intervality. These studies include Kohn's (1959) study of the marine gastropod genus Conus, and Hartley's (1949) study of river fish. In all of these, the diets of some predators have many more prey items in them than others, probably because these predators were more abundant and larger samples were easier to obtain. Similarly, some prey are encountered more often in the diets, probably for a variety of reasons, including abundance and palatability.

### Niche geometry and dietary opportunism

I first wish to consider how niches will appear if the predators take essentially the same species of prey. Superficially, the niche geometry would be rather dull. Any prey common to all of a set of predators implies that each predator overlaps with every other predator. The predator overlap graphs are both rigid circuit and trivially interval. Similarly, if all the predators exploit exactly the same set of prey, then the prey overlap graphs will be superimposed and there will be no holes.

Now I think that this complete overlap in resources is improbable if not impossible. But considerable overlap is clearly a strong tendency to be observed in the various studies I have examined. The question we must now ask is what will predator overlap graphs and assembled prey overlap graphs look like under this model of dietary opportunism given the inevitable sampling problems? I shall argue that the answer is "just like the ones we observe". There is a strong tendency towards rigid circuit and interval predator overlap graphs and assembled prey overlap graphs without holes.

## Niche overlap graphs

If the various predators essentially do take the same prey, then the common prey species will appear in the diets of more species of predators and the common predators will appear to take a wider variety of prey. Clearly, single records of prey species can only be in the diet of one predator and a single feeding record of a predator can be of only one prey type. For anyone unfamiliar with the data: yes, even the more extensive studies often have such poorly documented species in them. Single observations of prey are very common in these studies and 3 of the 6 quantitative studies in Cohen (1978) have predators with only one or two total prey items and all of them have predators with 20 or fewer prey records.

Now, recall that Cohen found that interval predator overlap graphs occurred more often than would be expected by chance alone. Cohen carefully created null models that generated the same number of predator overlaps as in the observed webs. But an abundant prey, common to all the predators' diets gives the maximum possible number of overlaps: every predator overlaps with every other predator. In such a case there is no variation in the null models against which to compare: real and model webs will be interval.

If the reality is that there is at least one prey species common to all predators, why do we ever find non–interval predator overlap graphs? The answer, I argue, lies in the large variations in predator and prey abundances — some overlaps will be missed because the sample sizes are too small to detect them. For example, if 10 predators actually share one common prey item then there would be 45 overlaps between the predators and, of course, the web would be trivially interval. There is only one way in which the niche overlap graph could be constructed. But if we missed 10 of those overlaps, there are a large number of ways in which we could construct the overlap graph and many of them would not be interval. Simply, it is possible that rigid circuit interval graphs are the 'reality'; observed patterns (some non–interval webs) deviate from this because of incomplete data. By having less than total overlap, variation in the positions of the overlaps between predators are possible in the null models.

An example of these points is provided by Kohn's study of the Conus gastropods on the subtidal reefs of Hawaii. This web has 13 species of predator and the overlap graph is not interval. Three of the predators have little overlap with the other 10, and these 10 have overlaps described in Table 1. Three species (1,2,6) overlap extensively with the other 7; it is in these 7 that an asteroid occurs (Fig. 3b) which destroys the intervality.

Is this pattern really surprising? What should we expect if all 10 predators had essentially the same choices of prey, (so the overlap graph would be trivially interval) and we had data which were more complete for some predators than others? We would find that those species with the largest sample sizes would appear to have more prey than those with smaller samples and, moreover, they would be more likely to overlap with more species. If we had few data on each of two species, then the chances of their overlapping would be slight. For this example, this pattern is obvious by inspection of the sample sizes in Table 1. For example, the smallest sample sizes are for predators 5 and 9 (6 and 1 prey observations respectively). They do not overlap in diet. If they did, the overlap pattern would no longer be asteroidal, and the web would be interval.

For this study, the overlaps between predators agree well with the view that 10 of the predators share prey extensively (and so the underlying overlap graph is rigid circuit and interval), but that inadequate data on rare species lead to our observing non–interval overlaps.

When food webs are reported in only binary form, detecting a lack of overlap because of inadequate data will be impossible. Yet, from the ways in which observers report food web interactions (discussed above) we should expect the effects to be the same. Poor sampling will make closely related predators appear less similar in diet than they really are.

### Holes in resource graphs

As already discussed, if predators take the same prey species, than their resource graphs will be super–imposed. The sampling problems I have discussed will lead to less apparent overlap. A common (= well–sampled) predator will appear to take a large variety of prey; predators sampled less well will appear to take smaller and different subsets of the common predator's diet. The assembled prey overlap graphs will take on the appearance of the petals of a flower — clusters joined together at the base by some prey items sufficiently common to be found in the diet of even the poorly sampled predators. Such clustering about a common prey is common among the resource graphs analyzed by Sugihara.

We might ask how species' diets differ from this model of dietary opportunism. An obvious test for this is the familiar $\chi^2$ contingency test. Certainly, diets are likely to differ from 'what we expect by chance'. The null hypothesis in this test is really the expectation that every predator has the same prey preferences. But we can also use the methodology to examine where the diets differ from this expectation. We can calculate what the expected number of a particular prey in the diet of a particular predator should be. This number is based on the knowledge of the total number of the given prey found in the diets of all the predators, and the total number of prey of all species in the diet of the given predator. We then subtract these expected numbers from the observed numbers. The positive values give us those prey species which appear more commonly than we would expect in the diet of a given predator. And from the positive deviations we can draw out the overlap graphs.

## TABLE 1

Overlaps between 10 species of predatory gastropod, studied by Kohn (1959). The species are arranged in decreasing order of number of feeding observations for each species.

Species

Number of Observations ↓ Species

| Number of Observations | Species | 2* | 3 | 6 | 1 | 4 | 8 | 10 | 7 | 5 | 9 |
|---|---|---|---|---|---|---|---|---|---|---|---|
| 58 | 2 | − | 1# | 1 | 1 | 1 | 1 | 1 | 1 | 1 | 1 |
| 40 | 3 |   | − | 1 | 1 | 1 | 0 | 0 | 0 | 1 | 0 |
| 36 | 6 |   |   | − | 1 | 1 | 1 | 0 | 1 | 1 | 1 |
| 27 | 1 |   |   |   | − | 1 | 1 | 0 | 1 | 1 | 1 |
| 16 | 4 |   |   |   |   | − | 0 | 0 | 0 | 0 | 0 |
| 15 | 8 |   |   |   |   |   | − | 1 | 1 | 1 | 0 |
| 15 | 10 |   |   |   |   |   |   | − | 0 | 0 | 0 |
| 13 | 7 |   |   |   |   |   |   |   | − | 1 | 1 |
| 6 | 5 |   |   |   |   |   |   |   |   | − | 0 |
| 1 | 9 |   |   |   |   |   |   |   |   |   | − |

*Key. All species are of the genus Conus: 1 sponsalis, 2, abbreviatus, 3, ebraeus, 4, chaldeus, 5, miles, 6, rattus, 7, distans, 8, vexillum, 9, vitulinus, 10, imperialis.

# Indicates species share at least one prey species.

Figure 4. A assembled prey overlap graph for fishes in the River Cam, using data from Hartley, (1949).

Unlike the graphs in Figure 3, this graph connects prey only if they occur more commonly in the predator diets than we would expect on the basis of the predators having identical diets. Thus predator E feeds on pay species 15 and 22 more than is to be expected by chance. This graph has several holes, e.g. between 21, 7, and 9. Numbering of predators and prey follows Cohen (1978) and Hartley's table.

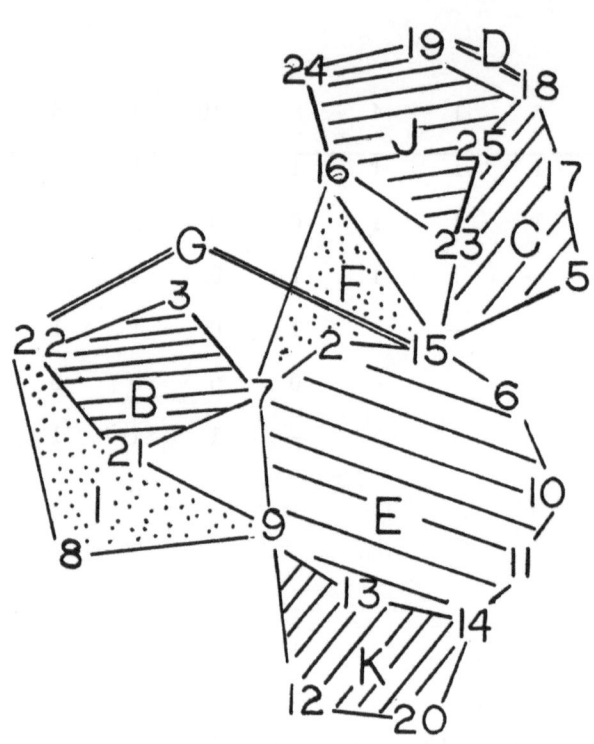

An example of this procedure, for Hartley's study of the fish in the River Cam is provided by Fig. 4. The original data show the assembled prey overlap graph to lack holes. When we look at only the positive deviations from what we would expect if all the species had identical diets, the situation is radically changed — the consumer graph has a considerable number of holes.

III.    THE GEOMETRY OF NICHES

I have argued that, given what we know, the topological patterns of intervality and rigid circuits and the scarcity of asteroids and holes, are not unexpected. These 'givens' include the strong tendency for taxonomically related predators in the same habitat to feed on similar prey and, furthermore, for some of the more abundant of those prey to appear in the diets of all of the predators. The 'givens' also include the fact that competitive pressures (and other factors) will cause the predators to differ systematically from this pattern of dietary opportunism. We add to those givens the inevitable errors of sampling which will cause us to underestimate the dietary overlaps between any pair of predators.

This tendency for dietary opportunism is hardly unknown. Most aquatic ecologists with whom I have talked have stressed the importance of chironomids in the diet of many freshwater fishes. It is the importance of chironomids that results in the patterns of almost total overlap in the study of the River Cam. We should expect similar results for other studies. Our failure to sample diets equally across predator species reduces the overlap we observe. This causes us to conclude, statistically, that the patterns in overlaps we observe are unusual.

At this point, it might appear that I am dismissing the patterns in the overlap graphs as just statistical artifacts stemming from analyzing familiar data with novel techniques. Nothing could be further from the truth. I argue that there are at least two interesting features in these results.

First, I suggest that the view of niches strung out along a single dimension is not a model which fits the preceding results data very well. Yet such a model will be very familiar to most ecologists even in its caricature of Fig. 5a. I suggest a better caricature is provided by Fig. 5b, where the niches are drawn to show (1) the extensive niche overlap formed by species exploiting some common prey species and (2) the resource partitioning which tends to give each predatory species a slightly different, though idiosyncratic, set of prey resources.

These two models are really extremes along a continuum. In (a) the niche centres are strung out in one dimension — though they could be across 2, 3...n dimensions. In (b) the niche centres are superimposed — they are essentially of 0 dimension. Which is the better model? I have argued that (b) is closer to the truth. What we really need to do is to quantify how strung out (1–dimensional) or clumped (0–dimensional) the niche centres of species groups really are.

Parenthetically, we should observe that this view of niche organization is probably identical to that suggested by Rosenzweig and Abramsky (1986). Working on desert rodents, they suggested that community organization might be best described as 'centrifugal'. They write: "If the primary preference of all species is shared, but their secondary preferences are distinct, (communities) are said to be organized centrifugally... Although the ideal combination of (habitats) is the same for many species in a guild, each species is adapted to tolerate relative deprivation of a different component of the mixture". Of course, Rosenzweig and Abramsky are concerned with the choice of habitats, rather than prey exploited, but one is likely to follow from the other, and, even if it does not, this convergence in species organization is certainly interesting.

For my second point, I return to a consideration of what motivated this chapter: how interactions are organized for <u>large</u> numbers of predators and prey. I have based by discussion on the organization of maybe a dozen species of predators. This is certainly more than the few species that predominate in most studies of community ecology, but it is far short of the hundreds or thousands of species that are found in real communities. How are these organized? What we know is that for small sets of species (perhaps a dozen or so, and perhaps representing what might be called guilds), we have diets that show a strong tendency to overlap and that this has implications for the topological patterns I have discussed. But how are these sets organized into the even larger sets of species—groupings that represent communities. And what topological features do these larger groupings possess?

It is quite possible that we do not know. The difficulties of getting quantitative data like those in the studies of Kohn (1959) and Hartley (1949) for large sets of species are all too obvious. What we have, then, are well—documented studies of a few species and much less satisfactory, non—quantitative data on the diets of much larger sets of species. The latter food web studies are highly aggregated. The components of these food webs are not species, but <u>trophic species</u> (Cohen, this volume) — groups of species which exploit the same (trophic) species and suffer predation from the same (trophic) species. Indeed, in their most aggregated form, many of the food webs analyzed in the literature contain such trophic species as 'plants', 'invertebrates', 'large ground animals', 'zooplankton' etc. Yet even among these webs, which are little more than quick sketches of real community organization, we find a scarcity of asteroids and holes and abundant interval, rigid circuit predator overlap graphs.

With this knowledge, there are really two possible kinds of organization that may describe large sets of species:

(a) Within small sets of species, overlap in diets is as I have suggested — considerable — and it accounts for the topological patterns that we observe in such groups. Across such sets of species, the patterns remain, but for different reasons. Thus, within the fishes in the River Cam, the predominance of chironomids in the diets ensures intervality in the predator

Figure 5. Two synoptic views of niche organization. In (a) there is the familiar
pattern of niches strung along some dimension — which might represent some
ordering say by size) of the prey species. The height of the lines represents the
frequency of use of a prey species. In (b) the niches are drawn in plan, and the
lines represent just one "contour" (i.e. frequency of use) of a two–dimensional array
of prey. The species depicted exploit very similar resources, all taking certain
abundant prey (at the centre of the cluster). Each species also selects some prey
uniquely.

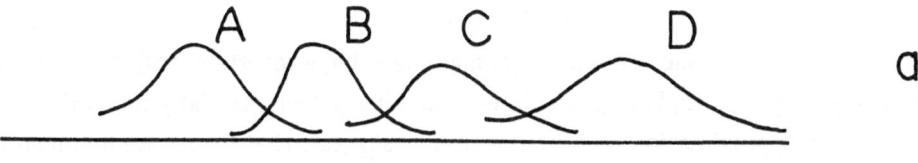

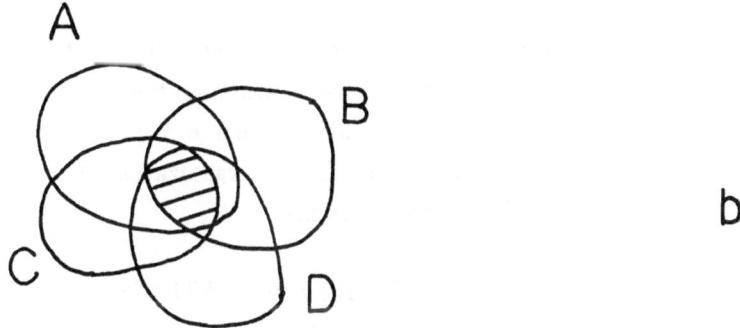

overlap graph. But across trophic species (say, fishes, aquatic birds, aquatic mammals, large invertebrate predators) the predator overlap graph might be interval because the fishes only overlap with birds, birds only overlap with mammals, and mammals only overlap with large invertebrates. If this is the case, we acknowledge we have an explanation for the patterns in prey use among a small number of species: they all share the same prey, because they are related species in the same habitat. But we still lack an explanation for the patterns among large sets of species, where the overlaps are strung out along some single, abstract, dimension.

(b) The second possibility argues that, at whatever organizational scale we view a community, the same patterns hold. Sets of species (or trophic species) show a tendency to exploit the same groups of species (or trophic species). Thus, the patterns of overlap between my hypothetical trophic species (fishes, birds, mammals, large invertebrates) in a river, would involve all the four groups exploiting say, insects, and hence being trivially interval.

To a rough approximation, the first possibility suggests that niches are 0–dimensional among small groups of species, but 1–dimensional among sets of small groups. The second possibility suggests that the dimension is always 0, irrespective of organizational scale. Cohen and Sugihara's studies show no evidence for higher dimensions (as typified by a non–rigid circuit niche overlap graph).

I do not know which of these possibilities is correct. Indeed, they represent ends of a continuum and there is no reason to think that all communities will be organized the same way. What these possibilities do tell us is that Cohen and Sugihara have provided us with a powerful tool for asking questions about community organization. Currently, we have neither the data nor the theory to answer those questions.

## IV.    THE PATTERNS:  ALTERNATIVE EXPLANATIONS

Sugihara (1984) has derived a simple assembly rule that is sufficient to explain the existence of rigid circuit predator overlap graphs, and the absence of holes in the assembled prey overlap graphs. The assembly rule, by its very nature, restricts the ways in which species can add on to a community. To understand the rule, we need another definition. Predators are considered adjacent if they share prey; thus, the predator overlap graph indicates which predators are adjacent. With this definition, Sugihara's rule is simple: predators can only add to adjacent predators. Is Sugihara's explanation different from the one I have proposed here? I think not. Rather the explanations are different facets of the same problem.

First some simple examples to illustrate Sugihara's ideas. Figure 6 shows which ways a predator may be added to a community and which ways are forbidden by Sugihara's rule. Before we ask why this rule works, consider the implications of examples fig 6 (a) and (e), which involve the same, final, predator overlap graph. One way of obtaining this community is permitted, the other is not. Thus, one sequence of species invasions is possible (1 and 3, then 4, and, finally 2); the other sequence is prohibited (1 and 2, then, 3; but we cannot then add 4). This means that we can obtain different communities (1, 2, 3, 4 or 1, 2, 3) depending on the sequence in which the species are added to the community.

Now, why does this rule work? Sugihara (1984) provides one explanation, by suggesting the "conventional wisdom" is that species tend to enter communities in order of decreasing polyphagy. If an incoming species of predator overlaps with two non–adjacent groups of predators it will likely be more polyphagous than any one of the predators with which it overlaps. But if the incoming species is relatively less polyphagous it will likely only overlap with adjacent predators. It is not at all clear to me that real communities work this way, or that it represents 'conventional wisdom'. If this is the mechanism, then why do species enter in such a strict sequence? Certainly, if this is the mechanism it proscribes a highly restricted sequence for species invasions.

Another way of viewing the assembly is to consider an example that, at face value, would favor a species breaking the assembly rule. In the food web at the bottom of figure 6 we see three predators feeding on a series of prey species, strung along some resource axis. To where should the next species of predator attach! Where there is least competition, of course. And, if the prey abundances are approximately equal, then the best place for the new predator to feed is at both ends of the resource axis: this is where the prey are being exploited by only one—and not two predators. This creates a non–interval, non–rigid circuit predator overlap graph, and an assembled prey overlap graph with a hole. We know this pattern of addition is prohibited in the real world. So what is wrong with my argument?

It could be that it is hard to do such different things as to feed on the ends of the resource axis. But many species are polyphagous. They might be able to exploit all the prey resources — but still only choose the ends because that is where there is least competition. Over long resource axes, from the ocean depths to the high Himalayas problem or from tiny herbivorous zooplankton to cows, it is clear why species do not choose the ends. But the data sets I have discussed (e.g. the Conus species) do not involve prey species that are over such long resource axes. They involve prey species that cannot by readily seen to be strung along some large, simple axis. Indeed, we know by the rarity of the prey ranking (the interval diet property) that prey are not ranked by predators like this.

So why is the pattern of invasion suggested by figure 6 difficult? Why is it hard to feed across the ends, when the "ends" are not, apparently special? Well, the "ends" are, in fact special, and for reasons to do with the dietary opportunism I discussed earlier. This same dietary opportunism creates a ranking of prey quite independent of any details of their natural history. Simply, some prey — even in models, but particularly in the real world — are much more abundant in the diet of predators than others. (These prey may be actually more abundant, or more available, or relatively more nutritious, or whatever. I shall talk about 'abundance' for simplicity.) These abundant prey species will be the ones selected first by the first predators to invade the community. The "ends" are simply the species that are rare and the "middle" of the resources are those that are common. Unless the effect of the predator already in the community is so very great that it depresses the common prey species below the abundance of the rare prey species in a community, predators will not create "donut" shaped holds in prey species abundances. Only if they did so would they force the later predators to enter the community by taking a selection of rare prey species. Predators may enter a community by taking rare species of prey, but they will also take the common ones too. Predators thus give the impression of entering the community and overlapping with over predators where there are more species competing for the resources, rather than where there are few species sharing those resources. In short, it is the range of prey abundances plus dietary opportunism that impose a ranking on the prey species, and this ranking gives the topological regularities.

This mechanism I have just outlined does not proscribe the strict ordered sequence suggested by Sugihara. A polyphagons predator species could enter the community early in the assembly or late — so long as it takes the common resources. But the resultant pattern will be the same as suggested by Sugihara's assembly rule. We will not observe predators feeding on an idiosyncratic selection of pay species and avoiding the central prey species.

The mechanism suggests some questions about whether a species can invade or not that are different from those usually asked. Is it the availability of rare prey species that enables a predator to invade, and so also exploit some of the common prey species? Or is it just the opposite: the availability of some left–over common prey may enable a species to enter the community and, incidentally, take some unexploited, rare prey species. We do not know the answer to these questions, because, to my knowledge, the problem of packing species into a community has never been framed in the context of this view of nich geometry before.

Figure 6. Rules for adding species to communities (partly after Sugihara, 1984). The top
figures are predator overlap graphs; circles are predators, lines connect predators
that overlap in diet; dashed lines and circles represent invading predators and their
interactions. Predator overlap graphs will be rigid circuit if invading predators
attach only to adjacent predators — i.e. predators connected directly. The
possibilities at the right are forbidden under this rule; those at the left, are
allowed. Note that the final graphs (e) and (c) are the same: one cannot be formed,
the other can, by the rules. This suggests that the sequence in which the species
invade may determine the end point. Species 2 can follow 1, 3 and 4, but species 4
cannot follow 2 and 3.

The bottom figure is a food web and corresponds to predator overlap graph diagram.
Why should the invasion suggested be forbidden, when the invading species takes
those prey species on which there is least predation?

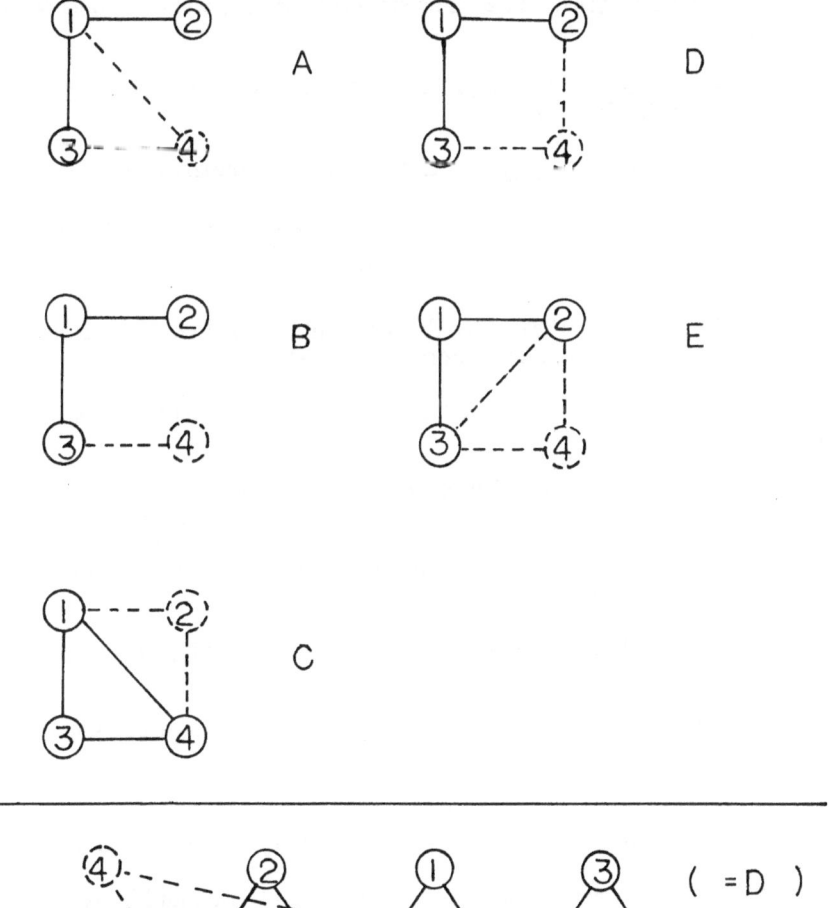

REFERENCES

Bird, R.P. 1930. Biotic communities of the aspen parkland of Central Canada. Ecology 11: 356–442.

Briand, F. 1983. Environmental control of food web structure. Ecology 64: 253–263.

Briand, F. and J.E. Cohen. 1984. Community food webs have scale–invariant structure. Nature 307: 254–267.

Cohen, J.E. 1977. Ratio of prey to predators in community food webs. Nature 270, 165–167.

Cohen, J.E. 1978. Food webs and niche space. Monographs in Population Biology 11, Princeton University Press, Princeton, N.J.

Cohen, J.E. and F. and Briand. 1984. Trophic links of community food webs. Proc. Natl. Acad. Sci. U.S.A. 81: 4105–4109.

Critchlow, R.E. and S.C. Stearns. 1982. The structure of food webs. Amer. Natur. 120: 478–499.

Hartley, P.H.T. 1949. Food and feeding relationships in a community of fresh–water fishes. J. Anim. Ecol. 17: 1–14.

Kohn, A.J. 1959. The ecology of Conus in Hawaii. Ecol. Monogr. 29: 47–90.

Lawton, J.H. 1989. Food webs. Symposium of the British Ecological Society.

Pimm, S.L. 1982. Food webs. Chapman and Hall, London.

Pimm, S.L. and R.L. Kitching. 1987. The determinants of food chain lengths. Blackwells Scientific Publications, Oxford, (in press).

Pimm, S.L., and J.H. Lawton. 1977. The number of trophic levels in ecological communities. Nature 268: 329–331.

Pimm, S.L., and J.H. Lawton. 1978. On feeding on more than one trophic level. Nature 275: 542–544.

Pimm, S.L., and J.H. Lawton. 1980. Are food webs divided into compartments? Journal of Animal Ecology 49: 879–898.

Rosenzweig, M.L. and Z. Abramsky. 1986. Centrifugal community organization. Oikos 46, 339–348.

Schoener, T.W. 1974. Resource partitioning in ecological communities. Science 185, 27–39.

Sugihara, G. 1982. Niche hierarchy: structure, organization, and assembly in natural communities. Ph.D. Dissertation, Princeton University, Princeton, N.J.

Sugihara, G. 1983. Holes in niche space: a derived assembly rule and its relation to intervality. In D.L. DeAngelis, W.M. Post, and G. Sugihara [eds.] Current trends in food web theory. Oak Ridge National Laboratory Technical Memoranda 5983.

Sugihara, G. 1984. Graph theory, homology and food webs. Proceedings of Symposia in Applied Mathematics 30: 83–101.

Yodzis, P. 1981. The structure of assembled communities. J. Theoret. Biol. 92: 103–117.

Yodzis, P. 1982. The compartmentation of real and assembled communities. Amer. Natur. 120: 551–570.

# CHAPTER 8

## The Dynamics of Highly Aggregated Models of Whole Communities

Peter Yodzis

Department of Zoology

University of Guelph

Guelph, Ontario

N1G 2W1  Canada

## I.    INTRODUCTION

I am going to use the term (ecological) <u>community</u> to mean the set of all living things in some given location.  Typically, a community in this sense will consist of hundreds of biospecies (even if we permit ourselves to neglect —— possibly at our peril (Ducklow et. al. 1986) —— microorganisms).  These are, then, very large, and maybe very complex, systems.

One way to deal with such large systems is to study only small parts of them at a time, that is, to study small subsets of species in isolation.  This is the approach of traditional ecological models (predator–prey, competition, and so forth), and of traditional ecological field study.  This approach is, indeed, so enshrined in custom that the term "community" is very frequently used in ecology to denote small, functionally (or taxonomically!) defined subsets of species.  Too little (other than Schaffer 1981, Bender et. al. 1984) has been done in the way of explicating in what way, and to what extent, it makes sense to study these kinds of systems as though they existed in isolation; but this approach has enjoyed considerable success, and has certainly shaped the myths and metaphors in terms of which we currently think about communities.  (Lest there be any misunderstanding: I am using the word "myth" here not in any pejorative sense, but in the classical sense of, say, Frankfort et. al. 1946.)

We have here a couple of questions of scale: scale in what we conceive of as "the system", and scale in how finely differentiated in "taxon space" our view of that system is.  In the traditional approach, the basic units of study are usually populations; in particular, organisms are usually distinguished right down to the level of biospecies.  "Community" is conceived of in a very constricted sense, but this object is studied with very high taxonomic resolution.  Another approach, obverse to the traditional one, is to study whole communities as I have defined the term above, but to do so with very low taxonomic resolution.  This is the approach that I am going to discuss here.

A major point of departure for this approach in recent years has been a concentration on the trophic aspect of species interactions, as exemplified by the idea of <u>food web</u>.  The construction of a food web corresponding to some real community consists of two steps: (i) lumping together of those biospecies with similar feeding habits into <u>trophospecies</u>, (ii) listing which trophospecies eat which others. If we carry out (i) and (ii) for all the species in a community (as I have defined the term), we get a <u>community food web</u>. Throughout the following I will use the term food web to mean community food web.

A serious problem with current food web data is the lack of a consistent usage for the term "similar" in (i). Most food webs in the literature were explicated by investigators who were really interested in some particular group of organisms, typically at high trophic levels, so that the aggregation in (i) is usually not even done consistently <u>within</u> food webs, much less among them. However, food web theorists, while aware of this problem, have chosen to ignore it for the time being, on the basis that it will be a very, very long time before a significantly better data base is built up, and in the meanwhile we would be foolish not to make what we can of the data we've got.

The "listing" in (ii) can take the form of a directed graph, such as Fig. 1, where each vertex represents a trophospecies and an arrow from vertex i to vertex j means that species j consumes species i. Fig. 1, which is a food web for Narragansett Bay, Rhode Island (Case 7, Briand 1983), is typical of the existing data, both in being highly aggregated and in being more highly aggregated at lower trophic levels.

A major virtue of food webs as an object of study is the existence of a standard compilation of community food web data, from a wide variety of habitats and geographic locations (Briand 1983). Indeed, ever since the first such collection (Cohen 1978), awareness among ecologists of the value of food web data has been growing, and we can expect bigger and, to some extent, better food web collections in the future.

I would like to discuss here some work that I have done on the dynamics of these highly aggregated representations of ecological communities. I will convey some good news and some bad news: the good news is that, in what I call a "quasiempirical" test, a very simple–minded dynamics for these already very simple–minded systems seems to make any kind of sense (Section 2). The bad news is that if so, then the outcomes of perturbation experiments on these systems are in significant measure indeterminate, when viewed on a relatively long temporal scale (Section 3).

II.    TROPHODYNAMICS AND PLAUSIBLE COMMUNITY MATRICES

It is very tempting to think of trophospecies as if they were populations, and to associate population–dynamical models with food webs. Thus, for a given food web associated with some ecological community, let $N_i$ be the density of trophospecies i, and assume an autonomous dynamics of the form

(1)                    $dN_i/dt = f_i(\mathbf{N}), \quad i=1,2,\ldots,s,$

where there are s trophospecies in the community, and $\mathbf{N}$ represents the vector

Figure 1.

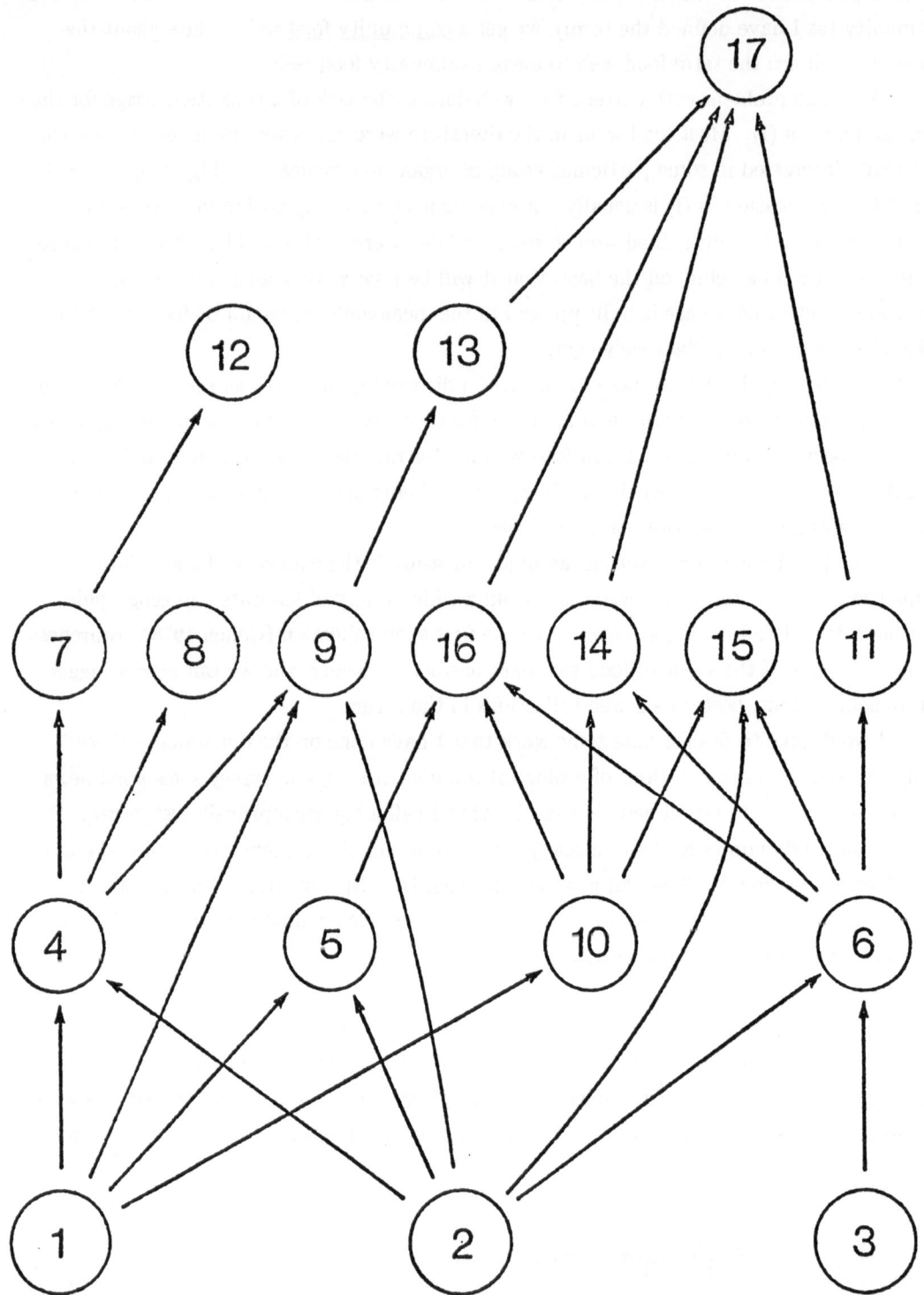

$(N_1, N_2, \ldots, N_s)$ of densities. One may call such a model <u>trophodynamical</u>.

Like any mathematical model of a "real" system, this one is a metaphor with no ultimate intrinsic justification. Its only justification is if it "works" when confronted with data from the "real" system it is meant to represent. In this section I will test, in a roundabout, quasiempirical way, the proposition that models of the form (1) can sensibly be associated, <u>via</u> food webs, with whole ecological communities.

Studying systems like this is, of course, different from studying Lotka–Volterra systems (despite the writings of some commentators). Lotka–Volterra models are of the form (1), with quadratic right–hand–sides, but I am not making any assumption about the functional forms in (1), except existence of a feasible equilibrium. The price I pay for this generality in the models themselves is that my analysis of these models can only be local in nature. Lotka–Volterra models, on the other hand, enable global analyses, but only under the very restrictive (global) assumption that the right–hand–sides in (1) are quadratic.

Assume for the time being that the system (1) has an equilibrium at some point $N_e$ in phase space, with $N_{ei}>0$ for all i. Then local stability of that equilibrium is determined by the eigenvalues of the <u>community matrix</u>

(2)
$$A_{ij}=(\partial f_i/\partial N_j) \Big|_{N_e}.$$

If we neglect all interactions other than intraspecific interference and trophic interactions, then we know the signs of all the matrix elements $A_{ij}$. Namely: (a) due to intraspecific interference, $A_{ii}<0$ for all i, (b) if species i is eaten by species j, then $A_{ij}<0$ and $A_{ji}>0$, (c) all other matrix elements are zero.

What can we say of the magnitudes of these matrix elements? Suppose we can "guess" these magnitudes to within an order of magnitude. Then we can view each $A_{ij}$ as chosen at random from an interval $\pm[B_{ij}/10, B_{ij}]$, and our problem is to find the $B_{ij}$. One might go so far as to argue that this is the best we could ever hope to do anyway, on two bases: (A) If we were to go out and literally try to measure all these quantities, it would be such an immense and difficult task that we could count ourselves lucky to achieve even this much accuracy. (B) Apart from any attempt to measure these parameters, due to the many sources of natural variability in real communities, these parameters might well not intrinsically possess exact values at all. It may only be meaningful to assert that their values lie in such–and–such intervals of real numbers.

Equation (2) tells us that $A_{ij}$ is the per capita effect of species j on the growth rate of species i. This enables us to make plausible guesses, based on the nature of the particular organisms involved, as to the relative magnitudes of some of the $B_{ij}$. For instance, if insects consume trees, then the per capita effect of insects on trees will be much smaller than the per capita effect of trees on insects. (The overall scale of the $B_{ij}$ is basically a choice of time units.)

I call a community matrix which is obtained in this way from data for some real community a plausible community matrix for that community (Yodzis 1981). A plausible community matrix reflects not only the topology of trophic interactions in the community, through the pattern of signs of its elements, but also the specific organisms in that community, through the pattern of magnitudes of those elements.

Now we can perform on a computer the following operation, for any given observed food web. First, generate 100 plausible community matrices as above. Make a note of the proportion of these that are stable. Then, disrupt each plausible community matrix into a new community matrix, by randomly permuting all positive off−diagonal entries among themselves, and all negative off−diagonal entries among themselves. Note the proportion of these "disrupted" community matrices that are stable.

If we do this for the 40 food webs in the collection of Briand (1983), we obtain Table 1 (Yodzis, 1981). Thereby we discover a very clear pattern: in every case where the plausible community matrices have a probability $\geq 0.01$ of stability, the disrupted community matrices have a smaller probability. Now, the disrupted community matrices have the same topological structure (food web) as the plausible ones, and the same average interaction strength. All that is different is that the detailed pattern of interaction strengths (magnitudes of the $A_{ij}$) in the disrupted matrices no longer corresponds to the particular organisms in the real community, as does this pattern in the plausible matrices. There is a very strong correlation between this pattern, and stability of the community matrix.

I think one is forced to conclude that either this correlation is a monumental coincidence, or the whole viewpoint embodied in the dynamics (1) (including a high degree of taxonomic aggregation) makes some kind of sense as a representation of these systems.

I have motivated this discussion by assuming a point equilibrium, and of course stable equilibria do have stable community matrices. When I first did these calculations (Yodzis 1981), I felt that they lent support not only to the form (1) for the dynamics, but also to the notion that these systems have point attractors. But what of other attractors? Have they any association with stable community matrices?

Consider a nonsingular point P in the phase space of a system of the form (1). Let x be an infinitesimal vector at P which is carried along by the flow associated with (1) (the "variation" in the sense of Nemytskii and Stepanov 1960). Let $A_{ij}$ be given by (2), with the evaluation taken at $N_P$ instead of at $N_e$. Then to first infinitesimal order,

$$(3) \qquad dx_i/dt = \sum_j A_{ij}x_j .$$

Table 1. Stability in plausible and disrupted community matrices.

| Case | Community | Fraction of community matrices which are stable | |
|------|-----------|---------|-----------|
| | | Plausible | Disrupted |
| 1 | Cochin estuary | .12 | .05 |
| 2 | Krysna estuary | .64 | 0. |
| 3 | Long Island estuary | .49 | 0. |
| 4 | California salt marsh | .52 | .03 |
| 5 | Georgia salt marsh | 1. | .22 |
| 6 | California tidal flat | .07 | 0. |
| 7 | Narragansett Bay | 0. | 0. |
| 8 | Bissel Cove Marsh | .07 | .02 |
| 9 | Lough Ine rapids | .43 | .03 |
| 10 | Exposed intertidal (New England) | .89 | .15 |
| 11 | Protected intertidal (New England) | .59 | .12 |
| 12 | Exposed intertidal (Washington State) | .41 | 0. |
| 13 | Protected intertidal (Washington State) | .19 | 0. |
| 14 | Mangrove swamp (Station 1) | .20 | .13 |
| 15 | Mangrove swamp (Sta. 2) | .24 | .09 |
| 16 | Pamlico River | .45 | 0. |

Table 1 (continued).

| | | | |
|---|---|---|---|
| 17 | Marshallese reefs | .39 | 0. |
| 18 | Kapingamarangi atoll | .33 | 0. |
| 19 | Moosehead Lake | .74 | 0. |
| 20 | Antarctic pack ice zone | .09 | 0. |
| 21 | Ross Sea | .12 | 0. |
| 22 | Bear Island | .01 | 0. |
| 23 | Canadian prairie | .19 | 0. |
| 24 | Canadian willow forest | .92 | .04 |
| 25 | Canadian aspen forest | .95 | .02 |
| 26 | Aspen parkland | .28 | 0. |
| 27 | Wytham Wood | .12 | 0. |
| 28 | New Zealand salt meadow | .78 | 0. |
| 29 | Arctic seas | .07 | 0. |
| 30 | Antarctic seas | .47 | 0. |
| 31 | Black Sea epiplankton | 0. | 0. |
| 32 | Black Sea bathyplankton | 0. | 0. |
| 33 | Crocodile Creek | .87 | 0. |
| 34 | River Clydach | .07 | 0. |
| 35 | Morgan's Creek | .71 | 0. |
| 36 | Mangrove swamp (Sta. 6) | .26 | 0. |
| 37 | California sublittoral | .76 | 0. |
| 38 | Lake Nyasa rocky shore | .63 | 0. |
| 39 | Lake Nyasa sandy shore | .67 | 0. |
| 40 | Malaysian rain forest | .34 | .02 |

Thus the community matrix A is a direct measure of the instantaneous expansion/contraction of the flow. For instance, if we set

(4) $$D=\left(\sum_i x_i x_i\right)^{1/2} ,$$

we find from (3)

(5) $$dD/dt=\sum A_{ij}x_ix_j/D .$$

Assuming A is semisimple (hence normal), and using its spectral decomposition in (5), one can show that

(6) $$dD/dt=\sum\sum Re(\lambda_a)(v_a)_i(v_a)^* ,$$

where $\lambda_a$ are the eigenvalues of A, $v_a$ are the corresponding eigenvectors, and $^*$ means complex conjugate.

What this says is that the sign of the real part of each eigenvalue determines whether there is instantaneous expansion or contraction in the corresponding eigendirection. This is essentially a local analog of the theory of Liapunov characteristic exponents (e.g., Nemytskii and Stepanov 1960). Note in particular that the component longitudinal to the flow (taking $x_i$ in the same direction as $dN_i/dt$) corresponds to instantaneous acceleration in the movement along the trajectory through P.

Now suppose the system has a stable cycle. What does this entail for the stability of A? For P on the cycle, there will be contraction in any direction transverse to the flow. So s−1 of the s eigenvalues will have negative real parts. Obviously, for any cycle the acceleration along the orbit must either be zero everywhere, or positive at some points of the orbit and negative at others. Rather clearly, the second case is generic. So almost always, when there is a stable cycle there will be points in phase space where A is a stable matrix.

It is difficult to generalize about chaotic attractors in this regard. There are points on the Lorenz attractor where A is stable, but not on the "spiral" attractor of Rössler. I would venture a guess that neither existence nor nonexistence of points where A is stable is a generic property of strange attractors.

Therefore, while I still feel that the results in Table 1 lend credence to the whole idea of associating dynamical systems of the form (1) with highly aggregated food webs, I am not so sure as I was when I wrote my 1981 paper that they lend much credence to the existence of point attractors for these systems.

III.    PERTURBATION EXPERIMENTS
Despite the conclusion of Section 2, it is still reasonable to study equilibrium dynamics

as an ansatz, and that is what I am going to do in this section, in order to bring out another aspect of scale in the study of ecological communities, namely the scale of time over which we make observations.

In particular, I am going to consider press perturbations (Bender et. al. 1984) of these systems. A press perturbation experiment on an equilibrium community has the following form. First, the densities are observed at equilibrium. Then, members of one or more species are added to or removed from the system, and this addition/removal continues at a fixed rate for the entire duration of the rest of the experiment. After the system has again equilibrated, the new densities are observed.

Press perturbations correspond to manipulations such as stocking or erecting exclusion cages or fences, which are commonly used in ecological field studies. Many unintentional perturbations, such as the introduction of toxins on a continuing basis, can also be viewed as press experiments.

These experiments involve both direct and indirect effects. Species $j$ has a direct effect on species $i$ if $A_{ij}$ is nonzero. Operationally, this means that if all densities other than $N_i$ and $N_j$ are somehow held constant and $N_j$ is altered, then $N_i$ will change.

Species $j$ has an indirect effect on species $i$ if there exists a sequence $\{k_1, k_2, \ldots, k_n\}$ ($n>0$, $k_a \neq i$ or $j$ for all $a$) of distinct species such that species $j$ directly affects species $k_1$, which directly affects species $k_2, \ldots$, which directly affects species $k_n$, which directly affects species $i$. Operationally, if species $j$ has an indirect effect on species $i$ but not a direct effect, and if all densities other than $N_i$ and $N_j$ are somehow held constant and $N_j$ is altered, then $N_i$ will not change; but if this same experiment is done without holding the other densities constant, then $N_i$ will change.

What about pure exploitative competition, which can be modelled either as an indirect effect, by explicitly including equations for resource dynamics, or as a direct effect, by using Lotka–Volterra equations with competition coefficients expressed through niche overlaps? Is it direct or indirect? I believe one has to say it is indirect. The direct Lotka–Volterra model is obtained from the more rigorous description with explicit resource dynamics only through an approximation (that the resource dynamics is very much faster than that of the consumers) which does not necessarily always hold (MacArthur 1972). The indirect model is both more fundamental (in the sense that the direct model can be derived from it, but not vice versa) and more generally valid.

The results of sufficiently small press perturbations on a community whose dynamics are given by a system of the form (1) are easily deduced. If we continually add members of species $j$ to the community at a rate $I_j$ members per unit area per unit time, the system (1) becomes

$$dN_j/dt = f_j(\mathbb{N}) + I_j$$

$$dN_i/dt = f_i(\mathbb{N}) , \quad i \neq j.$$

We have assumed the community was in equilibrium before the press. For sufficiently small $I_j$, there will still be a stable equilibrium, with densities $\mathbb{N}(I_j)$, and at this equilibrium we will have

$$0 = f_j(\mathbb{N}_e(I_j)) + I_j$$

$$0 = f_i(\mathbb{N}_e(I_j)) , \quad i \neq j.$$

Differentiating, we get

(7) $$\qquad\qquad dN_{ei}/dI_j = -(A^{-1})_{ij} .$$

So we can study the results of press experiments in the formalism of Section 2 by looking at the inverses of plausible community matrices. I will use the term effect of species $j$ on species $i$ to mean the influence of perturbations of species $j$ on the density of species $i$ in press addition experiments. (Of course, press removal experiments will yield the opposite influence.) According to (7), $-(A^{-1})_{ij}$ gives a numerical measure of the effect of species $j$ on species $i$. It is interesting to note that Hastings (1986) has found that the inverse matrix plays a central role also in the study of invasions.

The interaction strengths $A_{ij}$ in plausible community matrices are determined only to within an order of magnitude, so one expects a certain amount of random variation in the inverse matrices. What one finds is a great deal of variation. This variation is of three kinds: (1) variation in the strength of effects (i.e., in the magnitude of each element of $A^{-1}$), (2) variation in the direction of effects (i.e., in the sign of each element of $A^{-1}$), (3) variation in the topology of major effects (i.e., in the identity of those matrix elements which are the biggest).

I have studied this variation using Monte Carlo methods (Yodzis 1988). For instance, Table 2 summarizes directional indeterminacy in press perturbation experiments for plausible community matrices associated with 16 of the Briand food webs. I call the effect of species $j$ on species $i$ in some community directionally determined if the confidence level, in the statistical universe of plausible community matrices, for the sign of $-(A^{-1})_{ij}$ is at least 95%;

other wise I call it <u>directionally</u> <u>undetermined</u>. Table 2 shows that there is a great deal of directional indeterminacy in these effects. It also shows that, even when effects are directionally determined, it is not uncommon that they run counter to the seemingly most obvious expectations. For instance, one expects that adding predator individuals will have a negative effect on that predator's prey populations, but (Table 2) even in 11% of the <u>determined</u> predator–prey interactions in these communities, adding predators has a <u>positive</u> effect on the prey.

The results in Table 2 show that in these communities indirect effects are very frequently prevalent over direct effects.

One may well object that in this table strong and weak effects are mixed indiscriminately, and that while it is hardly surprising that weak effects should show this kind of sensitivity to the exact values of the interaction strengths, there is a good chance that the strongest effects (which are anyway the really interesting ones) will be more determinate. In fact, it turns out that this is not the case at all.

For any positive real number $x$, let $\mathrm{mag}\ x = 10^N$, where $N$ is the (unique) integer such that $10^N \leq x < 10^{N+1}$. For a given community matrix $A$, call the effect of species $j$ on species $i$ a <u>major effect on</u> $i$ if, for all $k$,

$$\mathrm{mag}|(A^{-1})_{ij}| \geq \mathrm{mag}|(A^{-1})_{ik}| \qquad ,$$

and say that it is a <u>major effect of</u> $j$ if, for all $k$,

$$\mathrm{mag}|(A^{-1})_{ij}| \geq \mathrm{mag}|(A^{-1})_{kj}| \quad .$$

Call the effect a <u>major effect</u> if either of these two conditions is satisfied. These definitions simply identify the major effects as those which are largest in order of magnitude.

What I have called topological indeterminacy has to do with variation in the identity of major effects. For instance, Fig. 2 and Fig. 3 depict, for two plausible community matrices associated with the Narragansett Bay food web of Fig. 1, the major effects on each species. Presence of a line from vertex $j$ to vertex $i$, with an arrowhead or circle at vertex $i$, indicates that the effect of $j$ on $i$ is major. If the line has an arrow (circle) at vertex $i$, the effect of $i$ on $j$ is positive (negative).

Clearly, the topology of major effects (in the sense of the topology of the corresponding directed graphs) can vary quite a lot. To quantify this, say that the effect of species $j$ on species $i$ is <u>unimportant</u> if, in a random sample of plausible community matrices constructed as in Section 2, the probability for it to be major is less than .05, and call it <u>most important</u> if this probability is greater than .95. The effects which are neither unimportant nor most

Table 2.     Directional indeterminacy in plausible community matrices,
categorized by type of effect.  N is the total number of effects in each category.

| Type | N | Proportion directionally undetermined | Among determined, proportion showing reverse effect |
|---|---|---|---|
| Self | 223 | .27 | 0. |
| Predator on prey | 317 | .52 | .11 |
| Prey on predator | 317 | .54 | .07 |
| Indirect | 3423 | .50 | — |
| Competitive | 566 | .58 | .29 |

Figure 2.

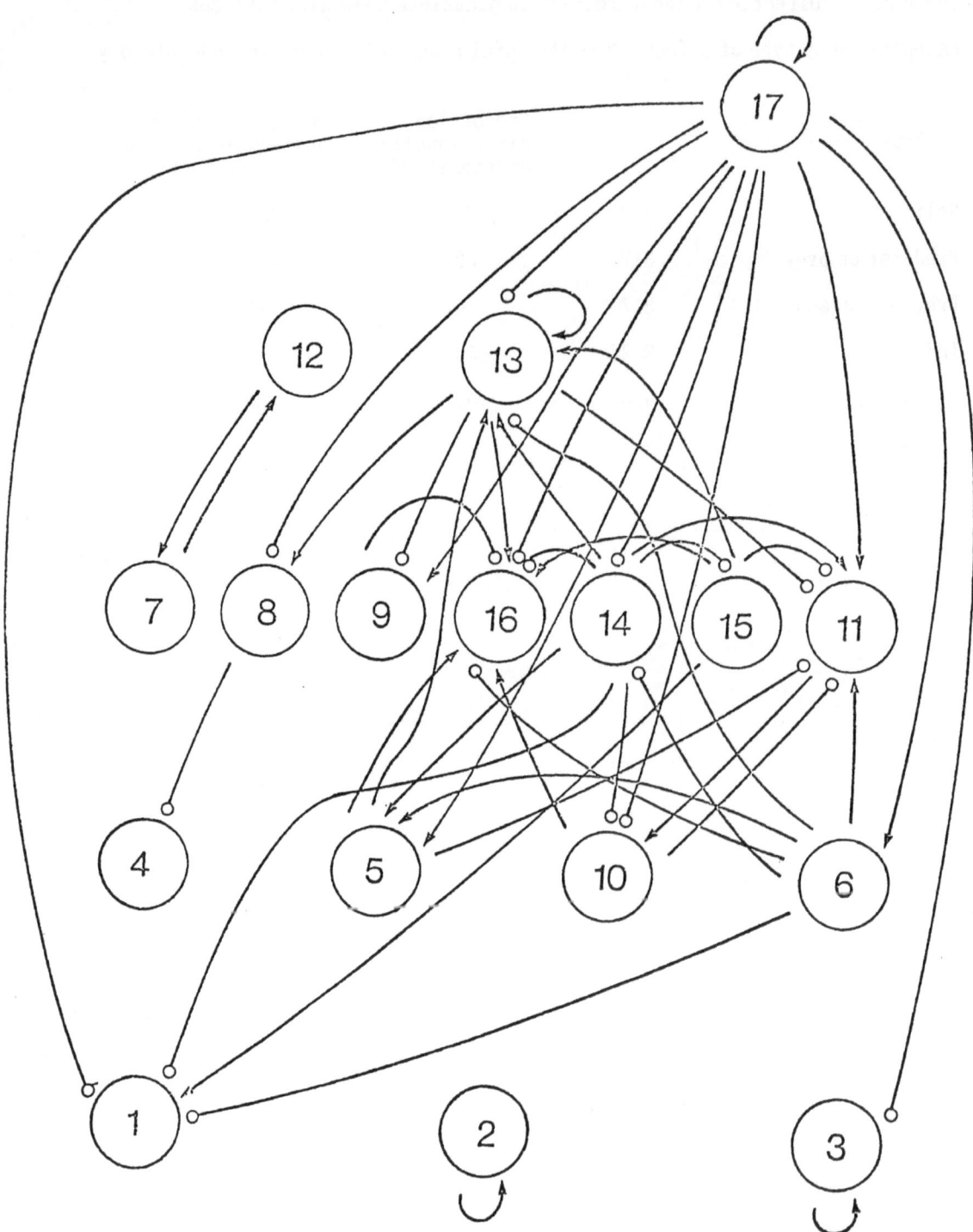

Figure 3.

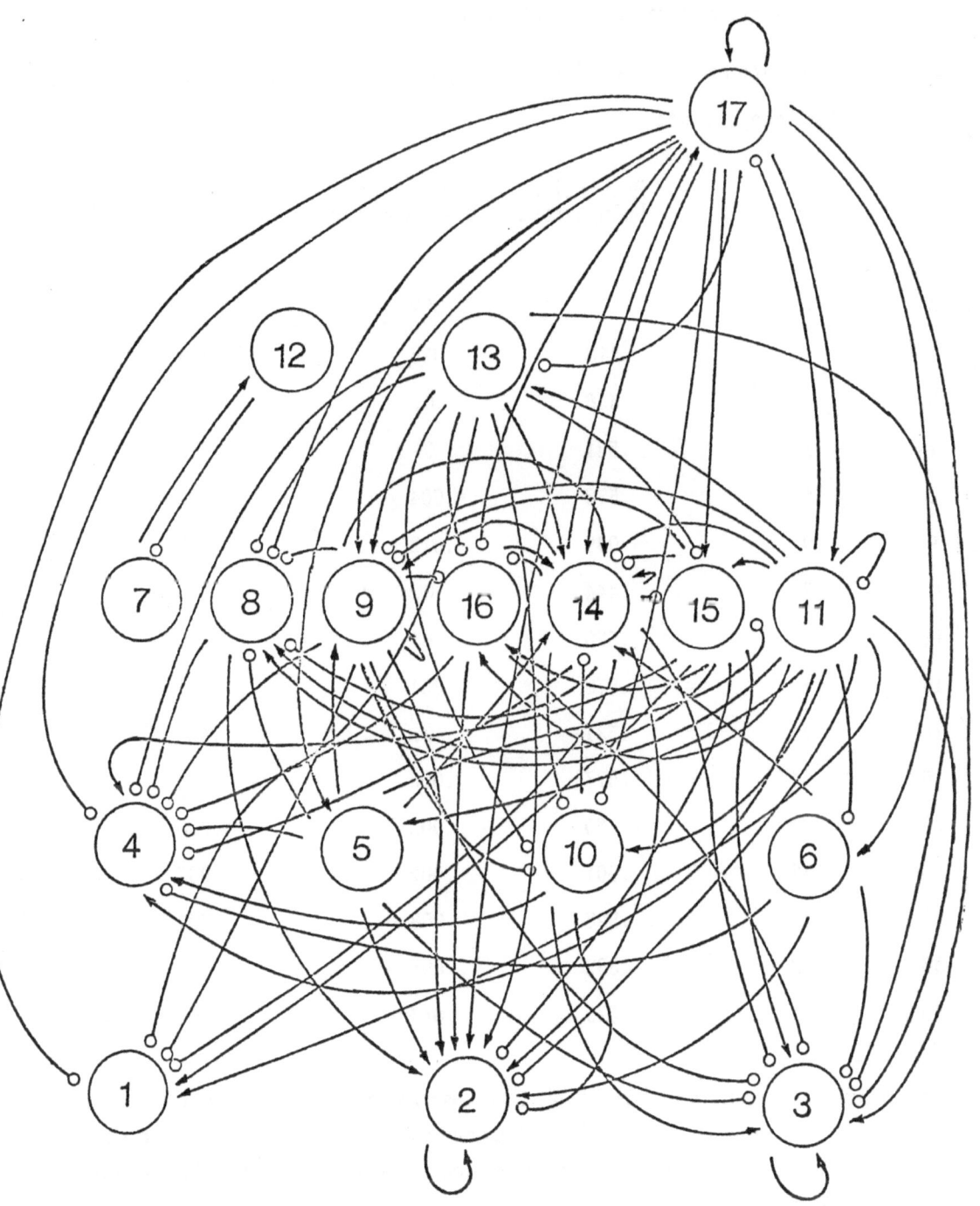

important are <u>topologically</u> <u>undetermined</u> in the sense that we cannot decide, at the 95%
confidence level, whether those effects will be major or not in a

Table 3. Proportion of effects which are unimportant, most important, and
topologically undetermined.

| Case | S | Proportion Unimportant | Proportion Most Important | Proportion Topologically undetermined |
|------|-----|------|------|------|
| 3 | 24 | .727 | .002 | .271 |
| 4 | 13 | .118 | .018 | .864 |
| 5 | 6 | .556 | 0. | .444 |
| 7 | 17 | .581 | 0. | .419 |
| 8 | 15 | .573 | .004 | .423 |
| 9 | 9 | .049 | .012 | .939 |
| 10 | 3 | .444 | 0. | .556 |
| 11 | 5 | .520 | 0. | .480 |
| 14 | 8 | .578 | .031 | .391 |
| 15 | 7 | .531 | 0. | .469 |
| 16 | 14 | .577 | .005 | .418 |
| 25 | 24 | .760 | .002 | .238 |
| 28 | 32 | .901 | .002 | .097 |
| 29 | 16 | .598 | .008 | .394 |
| 36 | 19 | .332 | .008 | .660 |
| 40 | 11 | .579 | .017 | .404 |

given plausible community matrix. Table 3 summarizes the proportion of effects which are
topologically undetermined in each of the 16 Briand communities considered. Surprisingly,
even the most important effects still show a considerable degree of directional indeterminacy,
as shown in Table 4.

I suggested in Section 2 that these highly aggregated representations seem to be not
entirely senseless. However, Table 3 suggests one possible shortcoming of aggregated models.

Table 4. Directional indeterminacy of most important effects in plausible community matrices, categorized by type of effect. N is the total number of effects in each category.

| Type | N | Proportion directionally undetermined | Among determined, proportion showing reverse effect |
|------|---|------|------|
| Self | 44 | .07 | 0. |
| Predator on prey | 55 | .31 | 0. |
| Prey on predator | 55 | .33 | 0. |
| Indirect | 769 | .30 | — |
| Competitive | 28 | .39 | .35 |

The number of trophospecies in these communities varies from 3 to 32. This variation bears little relation to the number of biospecies actually present; rather, it is due primarily to differing degrees of aggregation among the various studies involved. As shown in Fig. 4, there seems to be a trend for more highly aggregated models (smaller species richness) to have more topological indeterminacy. Topological indeterminacy may be to a large extent a pathology of highly aggregated models. This remark does not seem to extend to directional indeterminacy, however (Yodzis 1988).

I would like to make a final point having to do with scale, namely the temporal scale on which one makes these kinds of observations. Remember that the effects I have been talking about in this section have to do with the change in density, relative to the pre–press density, of each affected species after the system has reached a new equilibrium. The temporal evolution of the densities from old equilibrium values to new ("transient behavior" in the language of engineering) can be quite complicated.

The reason is that the time scale on which the different contributions to the effect operate will generally be different. In particular, it can happen that the indirect component of an effect operates on a longer time scale than the direct component. Then in short–term observations one will think that one is getting the (expected) direct effect; whereas if one had stayed around long enough, one would have seen the opposite effect in the long term.

Brown et. al. (1986) seem to have encountered this problem in their field studies of desert rodents. Figure 5 illustrates it using the Narragansett Bay community matrix corresponding to Fig. 2. Displayed in this figure is the transient behavior of the densities of

Figure 4.

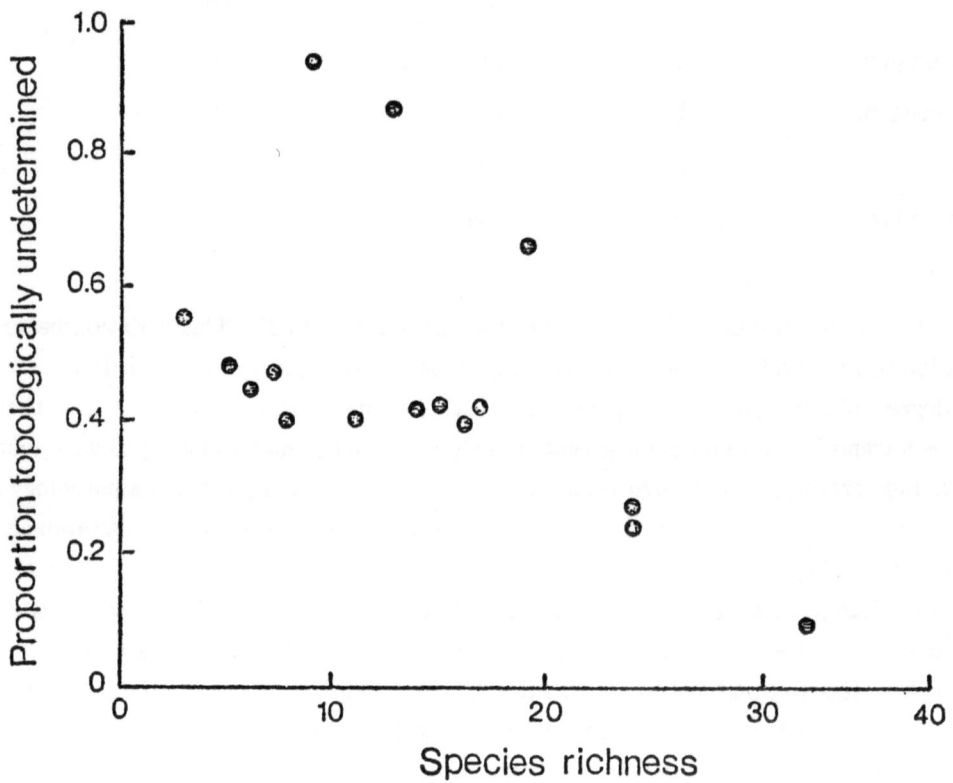

species 14 and 3, relative to their pre–press densities, after the onset of a continual removal of members of species 6. Short–term studies of environmental impacts are predicated on the assumption that all species behave like species 3 in Fig. 5; but species 14 does not behave this way.

These results suggest that it is at best extremely difficult to predict environmental impacts of human (or other) activities which can be viewed as press perturbations or (taken together with the findings of Hastings 1986) invasions.

This work was supported in part by the Natural Sciences and Engineering Research Council of Canada, under Grant No. A7775.

Figure 5.

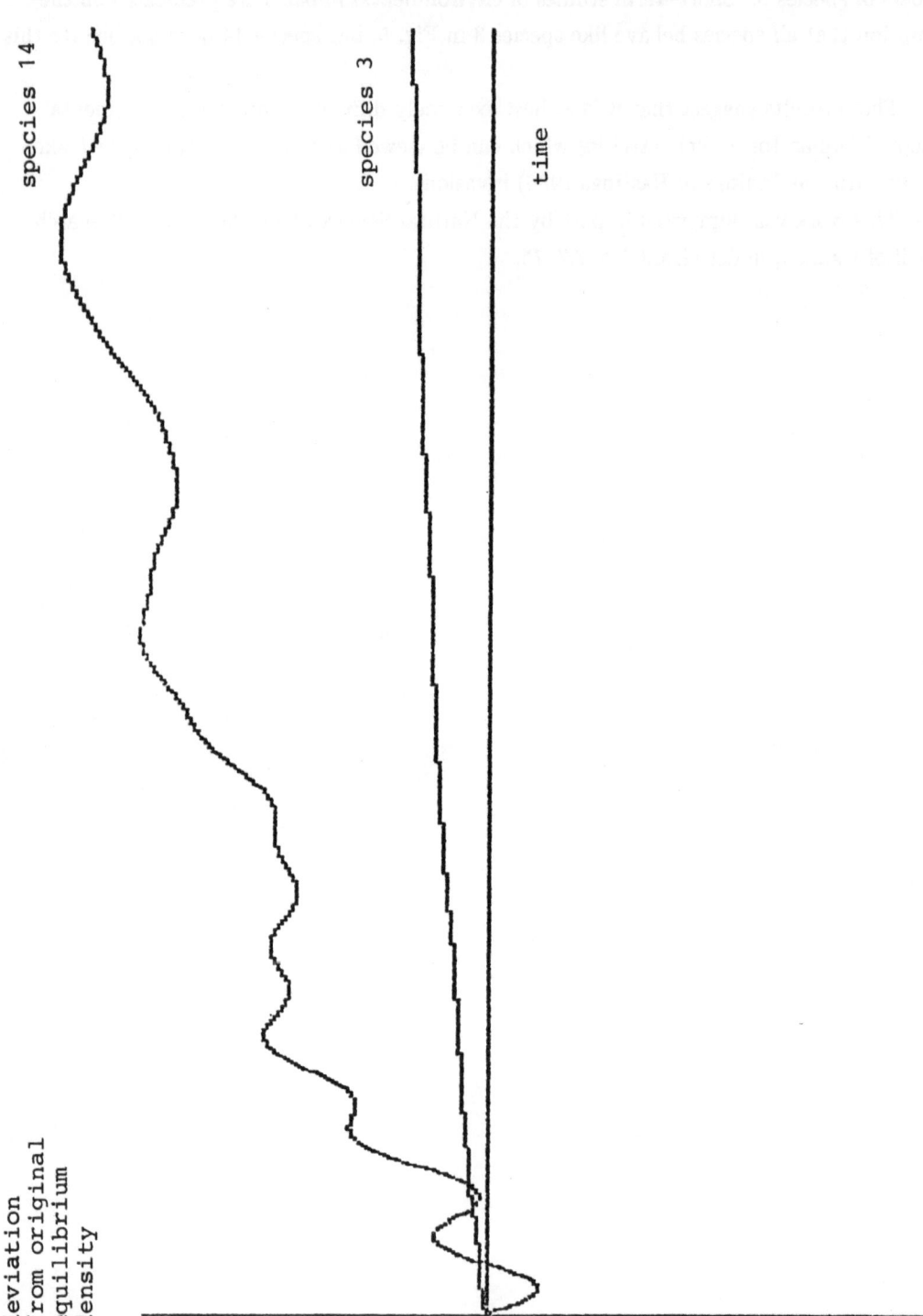

## REFERENCES

Bender, E. A., T. J. Case, and M. E. Gilpin. 1984. Perturbation experiments in community ecology: theory and practice. Ecology 65:1–13.

Briand, F. 1983. Environmental control of food web structure. Ecology 64:253–263.

Brown, J. H., D. W. Davidson, J. C. Munger, and R. S. Inouye. Experimental community ecology: the desert granivore system. In Diamond, J., and T. J. Case (eds.), Community ecology. Harper & Row, New York.

Cohen, J. E. 1978. Food webs and niche space. Princeton University Press, Princeton.

Ducklow, H. W., D. A. Purdie, P. J. LeB. Williams, and J. M. Davies. 1986. Bacterioplankton: a sink for carbon in a coastal marine plankton community. Science 232:865–867.

Frankfort, H., H. A. Frankfort, J. A. Wilson, and T. Jacobsen. 1946. The intellectual adventure of ancient man. University of Chicago Press, Chicago.

Hastings, A. 1986. The invasion question. J. Theoret. Biol., 121:211–220.

MacArthur, R. H. 1972. Geographical ecology. Harper & Row, New York.

Nemytskii, V. V., and V. V. Stepanov. 1960. Qualitative theory of differential equations. Princeton University press, Princeton.

Schaffer, W. M. 1981. Ecological abstraction: the consequences of reduced dimensionality in ecological models. Ecol. Monogr. 51:383–401.

Yodzis, P. 1981. The stability of real ecosystems. Nature 289:674–676.

Yodzis, P. 1988. The indeterminacy of ecological interactions as perceived through perturbation experiments. Ecology, in press.

*Your source for advances in theoretical biology and biomathematics*

Journal of

# Mathematical Biology

ISSN 0303-6812                    Title No. 285

**Editorial Board:** K. P. Hadeler, Tübingen; S. A. Levin, Ithaca (Managing Editors); H. T. Banks, Providence; J. D. Cowan, Chicago; J. Gani, Santa Barbara; F. C. Hoppenstedt, East Lansing; D. Ludwig, Vancouver; J. D. Murray, Oxford; T. Nagylaki, Chicago; L. A. Segel, Rehovot

**Subscription Information:**
To enter your subscription, or to request sample copies, contact Springer-Verlag, Dept. ZSW, Heidelberger Platz 3, D-1000 Berlin 33, W. Germany

For mathematicians and biologists working in a wide variety of fields – genetics, demography, ecology, neurobiology, epidemiology, morphogenesis, cell biology – **the Journal of Mathematical Biology** publishes:

- papers in which mathematics is used for a better understanding of biological phenomena
- mathematical papers inspired by biological research, and
- papers which yield new experimental data bearing on mathematical models

**Abstracted/Indexed in:** Current Contents, Excerpta Medica, Index Medicus, Mathematical Reviews, Science Abstracts, Animal Breeding Abstracts, Compumath, Helminthological Abstracts, Index to Scientific Reviews, Plant Breeding Abstracts, Zentralblatt für Mathematik

Springer-Verlag
Berlin Heidelberg New York
London Paris Tokyo Hong Kong

Springer

# Bio-mathematics

Managing Editor: S. A. Levin

Editorial Board: M. Arbib,
H. J. Bremermann, J. Cowan,
W. M. Hirsch, J. Karlin,
J. Keller, K. Krickeberg,
R. C. Lewontin, R. M. May,
J. D. Murray, A. Perelson,
T. Poggio, L. A. Segel

Springer-Verlag
Berlin Heidelberg New York
London Paris Tokyo Hong Kong

Springer